Russia and the West

How do Russian and Western perceptions of Arctic environmental affairs differ? Does the way we talk about the environment affect politics in the area?

The Russian part of the European Arctic contains some of the gravest environmental problems to be found in Europe today and this book discusses the East–West interface in Arctic environmental affairs since the early 1990s, tracing the dominant discourses in Russia and the Nordic countries.

Russia and the West traces similarities and differences between Russian and Western perceptions of these problems, of what causes them and of how they are being dealt with at the international level. It focuses on how environmental problems are framed and how this affects politics. Using a distinctive cross-cutting focus on environmental discourse and East–West relations, the author provides an in-depth analysis of the interface between Russia and Western countries over environmental issues such as nuclear safety, air pollution and the management of living marine resources.

Geir Hønneland is a political scientist and Director of the Polar and Russian Programme at the Fridtjof Nansen Institute, Norway. He has published widely on the management of natural resources and the environment in the European Arctic.

Environmental Politics/
Routledge Research in Environmental Politics
Formerly edited by Stephen C. Young

The fate of the planet is an issue of major concern to governments throughout the world and is likely to retain its hold on the agenda of national administrations due to international pressures. As an object of academic study, environmental politics is developing an increasingly high profile: there is great need for definition of the field and for a more comprehensive coverage of its concerns. This new series will provide definition and coverage, presenting books in the following broad categories:

- New social movements and green parties
- The making and implementations of environmental policy
- Green ideas

The series consists of two strands:
Environmental Politics addresses the needs of students and teachers, and the titles are published in hardback. Titles include:

Global Warming and Global Politics
Matthew Paterson

Politics and the Environment
James Connelly and Graham Smith

International Relations Theory and Ecological Thought towards Synthesis
Edited by Eric Lafferière and Peter Stoett

Planning Sustainability
Edited by Michael Kenny and James Meadowcroft

Routledge Research in Environmental Politics presents innovative new research intended for high-level specialist readership. These titles are published in hardback and include:

1 The Emergence of Ecological Modernisation
Integrating the environment and the economy?
Stephen C. Young

2 Ideas and Actions in the Green Movement
Brian Doherty

3 Russia and the West
Environmental co-operation and conflict
Geir Hønneland

Russia and the West
Environmental co-operation and conflict

Geir Hønneland

Routledge
Taylor & Francis Group

LONDON AND NEW YORK

First published 2003
by Routledge
2 Park Square, Milton Park, Abingdon, Oxfordshire OX14 4RN

Simultaneously published in the USA and Canada
by Routledge
711 Third Avenue, New York, NY 10017

Routledge is an imprint of the Taylor & Francis Group

First issued in paperback 2010

© 2003 Geir Hønneland

Typeset in Sabon by
BOOK NOW Ltd

British Library Cataloguing in Publication Data
A catalogue record for this book is available from the British Library

Library of Congress Cataloging in Publication Data
A catalog record for this book has been requested

ISBN13: 978-0-415-29835-3 (hbk)
ISBN13: 978-0-415-66627-5 (pbk)

Contents

Contents

Figures

Tables

Preface

This is a book I have wanted to write for a long time. As a student, researcher, interpreter and observer, I have been close for one and a half decades to the environmental interface between Russia and the West in the European Arctic. I have studied various aspects of Russian environmental management and East–West co-operation in this sphere from different practical angles and theoretical perspectives. However, I have not yet been able to pinpoint those social phenomena that follow from the fact that the issues at hand are simply 'talked about' differently in Russia and the West.

The research was made possible by two separate grants. The empirical data were mainly collected under the project 'Northwestern Russia as a Non-Military Threat to Norway: Mechanisms for Problem Solving at the International, National and Sub-National Level', financed by the Norwegian Ministry of Defence in the years 2000–2001. The Fridtjof Nansen Institute (FNI) provided internal funds that enabled me to finalise the manuscript during winter and spring 2002.

As usual, my former colleague and close friend Anne-Kristin Jørgensen has been my most important conversation partner on the themes of this book. We collaborated on several of the research projects that provided the empirical data for the study. Likewise, Arild Moe and Frode Nilssen worked with me on issues of nuclear safety and fisheries management. Bente Aasjord, Iver Neumann, Elena Nikitina, Ivan Safranchuk, Steven Sawhill and David Scrivener commented on various parts of the manuscript. Several of my FNI colleagues offered opinions and advice at an internal seminar on discourse analysis. Jildou Dorenbos and Lars Gulbrandsen went through several volumes of different Norwegian newspapers in search of articles on relations with Russia. Christel Elvestad, Christen Mordal and Morten Vikeby also provided assistance in the data collection phase. Thanks to you all.

Thanks also to the entire FNI staff, as well as our extramural assistants, for efficient support. In particular, I wish to extend my gratitude to my 'standing team' of assistants from recent book publications: to Chris Saunders for being a language consultant in the true sense of the word, to Maryanne Rygg for her professional technical assistance, and to Claes Rygge Ragner for producing all kinds of detailed maps on request.

In the transcription of Russian words into Latin characters, I have tried to pay attention both to general practice and consistency. Although I would have preferred to give priority to consistency, I have occasionally allowed exceptions in order not to depart from what might be considered general practice. Russian *e* is written as *ye* at the beginning of words and after vowels. Nevertheless, the *y* is skipped in proper names that already have a common spelling in English, e.g. Karelia. The Russian hard and soft signs are not transcribed.

All translations from the Russian and Norwegian are mine. The book contains many extracts from oral interviews and media reports. While it has been an objective to render these in correct English, I have also attempted to retain some of the original flavour of these mostly informal utterances in Russian and Norwegian.

Geir Hønneland
Lysaker

1 The study of environmental discourse

Introduction

This is a book about how we talk about the environment, why we talk about the environment in a certain way, and some of the effects of doing so. The study of environmental discourse is one of many ways of approaching the relationship between humans and the environment.[1] It may not be the most prolific so far in terms of empirical richness or theoretical maturity, but it has gained considerable momentum during the 1990s, drawing to a large extent on more comprehensive insights from other parts of the social sciences than those traditionally preoccupied with environmental politics. Discourse analysis has a particular focus on written and spoken language and seeks to trace patterns of how a given object of study is talked and written about by different subjects. Moreover, it is often a major goal of this approach to reveal the societal features, for instance values prevailing among a specific group of people at a given time, that produce particular verbal or written representations of reality. In the study of environmental politics, this often takes the form of tracing the context of this particular sector of policy; is the environmental discourse, for instance, embedded in other, and more general discourses in society? Finally, the question is sometimes asked whether tendencies to talk (or write) about the environment in a specific way contribute to shaping actual policy choices. For instance, does taking a conservationist stance in discussions lead to conservationist politics, or are there other motives behind the chosen form of expression besides a signalling of desire for particular political measures?

This is also a book about Russia and its relations with the West. Management of natural resources and the environment,[2] in recent years, has become one the most important institutional interfaces between East and West in the European Arctic.[3] This is commonly held to be a result of increased awareness of environmental issues in general as well as the obstructive environmental legacy of the former Soviet Union becoming ever more apparent towards the end of the 1980s. Russia has, on the one hand, displayed an increasing willingness to participate in international co-operative arrangements on management of natural resources and the environment. On the other hand, the Russians, due to financial constraints, have been 'forced' to accept foreign assistance to clean up their environmental disasters to an extent that was unthinkable in the late 1980s. The main theme of this book is how these interfaces or linkages between Russia and the West in the environmental sphere are spoken and written about in Russia and the West, respectively. How are the limits defined of what is to be perceived as legitimate knowledge, actor interests and institutional arrangements? Are there any differences between the predominant discourses on these issues in the Russian and Western part of the European Arctic? How are environmental discourses embedded in more overarching discourses in society in Russia and the West? Management of marine living resources, nuclear safety and industrial pollution are used as the primary empirical cases in the study.

If we were asked to characterise the eastern part of the European Arctic (see note 3) in one or two phrases, we would have to highlight not only its extremely bountiful natural resources but also grave environmental problems, particularly in the Russian part of the region. The area owes the existence of its human settlements largely to the extraction of natural resources. In the southern parts of the region, mainly in Sweden and Finland and Arkhangelsk Oblast of the Russian Federation, forestry has for centuries constituted the foundation for life.[4] In the more barren Murmansk Oblast in Russia, which geographically corresponds to the Kola Peninsula, fisheries and mining provided the industrial starting point for the build-up of large human settlements after the First World War, rendering the region the most densely populated area of the Circumpolar Arctic during the second half of the twentieth century.[5] The fishing grounds of the adjacent Barents Sea are among the most productive in the world, and the mineral deposits of the Kola Peninsula, mainly iron ore, nickel and apatite, are remarkable for their richness.[6] From the 1920s onwards, massive fishing fleets were built up in the region and, at the time of the break-up of the Soviet Union, Murmansk had the largest fish-processing factory of the entire Union.[7] Town names such as Nikel and Apatity, for their part, indicate the importance of the mining and metallurgical complex of the region.

However, the extraction of natural resources and the accompanying military build-up have taken place at the expense of the environment. Throughout the 1990s, Northwestern Russia has become more renowned for its environmental degradation than for its abundant resources.[8] Ever since Western journalists gradually gained easier access to this heavily militarised region from the mid-1980s, the black tree stumps of the dying forests around Nikel and Monchegorsk came for many in the West to symbolise the calamitous state of the environment in Russia. The nickel smelters of these two towns had virtually killed off the forests surrounding them and represented sources of pollution also for the neighbouring Nordic countries and other parts of Russia. 'Stop the death clouds!' was the slogan of environmental organisations in the Nordic countries in the early 1990s. The Nordic countries planned enormous assistance programmes to clean up production processes in the mining and metallurgical complex of North-western Russia, but nothing has come of these plans so far, mainly due to a lack of financial contributions and reliable clean-up schemes from the Russians themselves.[9] Financial hardships have forced the factories to reduce their activities in recent years, but Russian air pollution is still alarmingly high in the European Arctic (see Figure 1.1).

Throughout the 1990s, another environmental threat in the region upstaged air pollution as a focus of public concern, namely the danger of radiation from nuclear installations, discarded nuclear vessels, radioactive waste and spent nuclear fuel. The fire on board the Russian Northern Fleet's nuclear submarine *Komsomolets* and subsequent sinking of the vessel southwest of Bear Island in the Barents Sea in April 1989 was a rude awakening for the European public to the danger of nuclear radiation from nuclear-powered vessels stationed in Northwestern Russia.[10] Towards the end of 1990, rumours emerged that the Soviet Union had dumped radio-active material in the Barents and Kara Seas.[11] The rumours were officially confirmed in a Russian parliamentary report in the early 1990s (Yablokov *et al.* 1993). A major problem in the latter half of the 1990s was the build-up of radioactive waste and spent nuclear fuel in Northwestern Russia. Existing storage facilities were full, and no safe vehicles were available to transport the radioactive material out of the region for reprocessing or permanent storage. Moreover, financial problems forced the Northern Fleet to de-commission large quantities of nuclear-powered vessels in recent years.[12] Rumours also circulate about unsafe functioning of vessels still in service (Hønneland and Jørgensen 1999a). The *Kursk* accident in August 2000, albeit mainly a human tragedy, was a reminder of the potential dangers to the

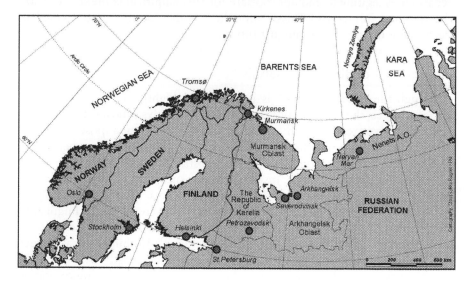

Figure 1.1 The European Arctic as defined in this book.

environment of the Northern Fleet's nuclear-powered vessels. Although radiation levels in the region are low at present, there are hazards associated with unsatisfactory storage of radioactive waste, decommissioned nuclear submarines awaiting dismantling and the continued operation of unsafe nuclear power installations, notably the Kola nuclear power plant at Polyarnye Zori.

Finally, signs of resource depletion have recently become evident in the region, most notably in the Barents Sea fisheries.[13] These fisheries, managed bilaterally by Russia and Norway since the mid-1970s, had for many years been seen as a management success. At the turn of the millennium, however, the Northeast Arctic cod stock, by far the commercially most important species in the area, appears to be in severe crisis. Some would have it that the situation is similar for the management system itself due to the dire state of its main object of regulation. There are indeed reasons for such an allegation: scientists are uncertain as to the size of the stock; managers do not follow the advice of the scientists in the establishment of quotas; and the enforcement system, at least on the Russian side, seems inadequate to keep track of actual catch levels and avoid fishing of juvenile specimens.

Among the empirical questions to be studied in this book are the following: what counts as legitimate knowledge for establishing quotas for marine living resources in Arctic ocean areas and for defining 'environmental problems' in relation to nuclear safety and air pollution? Why have these particular institutional arrangements been set up to combat environmental problems in the area? How can one explain the poor results in recent years of institutions responsible for managing marine living resources? What are the consequences of the fact that certain types of knowledge and institutions are regarded as legitimate and appropriate for this empirical context? How are the national interests of Russia and the Nordic countries intertwined in discussions of knowledge and institution?

The concept of discourse analysis

Although the term 'discourse' is employed in highly varying ways in the social sciences, the modern concept of discourse analysis most often seems to refer to a type of research activity carried out from a post-modernist viewpoint, aimed at revealing the influence of language on human behaviour and society. Typically, studies of discourses start out with a reference to the 'origins' of discourse analysis. Just as typically, such lines of reasoning end up with defining the works of Michel Foucault as a main epistemological inspiration of the discipline.[14] Before ending up there, debts to linguistics, post-structuralist philosophy, post-positivist philosophy of science or cognitive psychology are normally acknowledged.[15] Within political science and sociology, reference is made, *inter alia*, to reflectivist,[16] and cognitive approaches,[17] social constructivist theory,[18] post-positivist interpretative theory[19] and the critical theory of the Frankfurt school.[20] The main influence of Foucault (see, e.g., Foucault 1972, 1973, 1980) on discourse analysis is his insistence that power is omnipresent, and that power and language are inseparable. According to Foucault, individuals are always subject to the power relations – or discourses – in which they move and are never able to step back and make comparative assessments across various discourses. Foucault focuses on the subtle workings of power rather than on more visible expressions such as repression or domination, studying social relations to which the discourse of power is seldom applied, for instance schools and hospitals. He claims that power need not embody only negative qualities; what he refers to as 'the productivity of power' entails the construction of knowledge and discourse: '[power] traverses and produces things, it induces pleasure, forms knowledge, produces discourse' (Foucault 1980: 119, cited in Litfin 1994: 20).

What, then, is a discourse, a word that in everyday speech is used to denote a discussion or a conversation? In his textbook on discourse analysis, Neumann (2001: 17) lists a number of definitions of the term, describing discourse as '*a process* reflecting a distribution of knowledge, authority, and social relationships, which propels those enrolled in it',[21] '*a system* for the formation of statements',[22] and '*practices* that systematically form the objects of which they speak' (emphases added).[23] Add to the list definitions that present a discourse as '*the totality of things written and read, spoken and heard*' (Myerson and Rydin 1996: 4, emphasis added), '*a shared way of apprehending the world*' (Dryzek 1997: 8, emphasis added) and 'a specific *ensemble of ideas, concepts, and categorizations* that are produced, reproduced, and transformed in a particular set of practices and through which meaning is given to physical and social realities' (Hajer 1995: 44, emphasis added). Hence, discourse is perceived of as, *inter alia*, a system that produces something, a process of something being produced or reflected, practices that either produce or reflect something, that which is produced or reflected, or frames of viewing the world. More important than this divergence in categorising the discourse as a source or reflection of something is the conception of what this 'something' contains and what are its effects. Discourses are most often thought to produce or reflect specific ideas, concepts or statements.[24] These, in turn, are believed to affect those producing (or reflecting) them or their context; as mentioned above, they 'propel those enrolled in it', 'form the objects of which they speak' and 'give meaning to physical and social realities'. Hence, discourses are 'something' that affects the way 'someone' conceives of and talks about 'something'. Discourses can, in line with Foucault's conception of the intertwining of discourse and power, be regarded as practices or networks (or whatever) that compel subjects to view and speak of the world in specific ways. Not all subscribe to the notion of discourse and power as inseparable from each other, but the main point here is that subjects – as a result of power use or not – are at any time entangled in discourses that define the limits of how it is acceptable or 'natural' to perceive and refer to the physical and social environment.

As far as definitions are concerned, I adhere to the one that Litfin (1994) employs in one of the major works on environmental discourse. In her view, discourses are 'sets of linguistic practices and rhetorical strategies embedded in a network of social relations' (Litfin 1994: 3). This definition emphasises the *linguistic* and *rhetorical* aspects of that 'something' going on. Also, it links this linguistic 'something' specifically to a social 'something'.[25] A main premise of social science discourse analysis is the assumption that linguistic practices have something to say about the social world (which they either reflect or contribute to shaping). In line with this assumption, a major objective of this type of discourse analysis becomes to reveal what 'linguistic practices and rhetorical strategies' have to tell us about the social world. In general, discourse analysis has an *epistemological* rather than *ontological* objective. Hence, it does not primarily aim at telling us about the world as it *is* – according to Neumann (2001: 179), it is a basic ontological premise of the discourse analyst that the world presents itself to us as being more or less in a *flux* – but rather it aims at revealing *how our understanding of the world is created, maintained and reproduced*. The challenge lies in pointing to the social 'something' that brings forth, affirms and preserves specific modes of understanding. Some would also be concerned about the consequences of particular modes prevailing at particular times.

How does discourse analysis go about revealing the 'something' it decides to study? As Neumann (2001) notes, the declared inspiration of discourse analysis from post-positivist philosophy of science – compare the device of Feyerabend's (1975) classical attack on established thinking of method-ology: *anything goes!* – has prevented the elaboration of a methodology for discourse analysis. Neumann's (2001) own proposed method for discourse analysis includes a research process in three main steps:

- selection and demarcation of discourses;
- definition of discourse *representations*;
- definition of the discourse *stratification*.

The first step involves the selection of specific discourses to be studies and their delimitation in time. Neumann mentions the historical method of *genealogy* as particularly fruitful for discussing present events in terms of the past. Genealogy examines a particular phenomenon as a discursive practice. Focus is aimed at those conditions that enable the particular practice, at the practice itself and its effects. The method is an alternative to streamlined narratives of events, seeking to disclose the *contingency* of the process that has led to the contemporary situation. The aim is to 'trace every nook and cranny, developments that resulted in nothing, while showing how other ways of doing things are established, how every order carries within itself memories of other orders' (Neumann 2001: 153). Discourses consist of *representations* grouped together in specific ways. Representations are phenomena in the form they present themselves to us, i.e. not the things themselves, but filtered through what comes between us and the world (e.g. language and interpretative categories). Representations have to be reproduced (or *re-presented*) again and again in order to maintain their validity in the discourse of which they form part. It is the aim of discourse analysis to trace the position of various representations in the discourse to be studied. Finally, defining the stratification of the discourse involves examining whether all traits of a given representation are equally constant. It is reasonable to assume that unifying features of the discourse are more constant than those differentiating it. Moreover, material traits of a discourse are more difficult to 'explain away' than non-material ones.

This is only one example, and a very superficial reproduction of it indeed, of how to carry out discourse analysis. More flesh to the bone will be provided in the next section, which surveys some of the major contributions to the field from studies of the environment.

Three approaches to the study of environmental discourse

As mentioned in the introduction, the relationship between humans and the environment (understood as the physical environment) can be studied in a variety of ways, of which discourse analysis is only one – and so far a rather peripheral one. Ranging from law to economics and the 'softer' social sciences such as ethnology and social anthropology, the interest in human dimensions of environmental issues has been on the increase for several decades. Within political science, environmental politics have been studied from the macro-level of international relations to micro-level bureaucratic processes. Likewise, the methodological approach and epistemological heuristics of the studies vary to a high degree. It is not my opinion that discourse analysis should replace much of the ongoing research; there is little reason to perceive it as particularly 'better' than most other theoretical approaches (implying the rather pragmatic belief on my part that no theoretical approach can embrace all relevant aspects of a social phenomenon). However, I do subscribe to some of the fundamental ontological and epistemological claims of much discourse analysis – at least those found in the 'milder' variants prevalent in the study of environmental discourses – and I definitely think it has something of value to contribute that most other approaches lack. This will be further elaborated below.

The study of environmental discourse is not very widespread as yet. Moreover, the literature that exists is highly diverse in empirical scope, theoretical ambition and methodological refinement. Some works with titles indicating analysis of 'environmental discourse' do not live up to the expectations of those interested in this particular type of theoretical investigation.[26] The term 'discourse' is here often used in a more everyday sense similar to 'discussion' although such works occasionally refer to the theoretical debate on discourses. Other works flash Foucault and the rest of the post-structuralist classics, but empirically consist of rather 'traditional' analysis of political processes. A few works stand out in their theoretical clarity and empirical richness. This section presents three works that I have found particularly useful as a backdrop to my own study: Litfin's (1994) analysis of the formation of the global ozone regime, Hajer's (1995) on ecological modernisation in the Netherlands and the United Kingdom (with a main empirical focus on acid rain regulation) and Dryzek's (1997) discussion of the 'grand' global discourses on the environment.[27] These three books vary considerably with regard to both theoretical approach and empirical orientation, but all provide interesting input as to how discourse analysis can be performed in practice. A brief outline of these works is given below, with an emphasis on their theoretical focus rather than empirical findings. Their relevance for my own study is discussed in the subsequent section.

Litfin: trans-scientific problems and knowledge brokerage

In her much-cited and highly acclaimed study on the formation of the global ozone regime, Karen Litfin (1994) focuses particularly on the relationship between science and politics.[28] Placing herself in the context of a *reflectivist* approach to environmental politics (see note 16), in which she includes the epistemic communities literature, studies of social learning processes and negotiation-analytic modelling, she criticises the existing contributions to this literature for ignoring the relational aspects of science's influence on politics. For instance, she claims that the epistemic communities literature (see, e.g., Haas 1989, 1992a) views science as standing outside of politics and knowledge as divorced from power. According to Litfin, it suffers from a simplistic assumption of scientific consensus necessarily leading to political consensus. She views it as part of the 'rationality project' of 'rescu[ing] public policy from the irrationalities and indignities of politics' (Stone 1988: 4; cited in Litfin 1994: 4). She claims that, contrary to the belief in a sharp dividing line between science and politics, political arguments are often formulated in scientific terms. Questions about values are phrased as questions about facts. Facts, in turn, have to be expressed in language and are subjected to inter-pretation. 'Facts' are then chosen selectively by actors depending on their particular interests and audience.[29] *Knowledge brokers* – typically operating at low or middle levels of governments or international organisations – function as mediators between the original producers of the scientific knowledge and the political actors who use that knowledge but have neither the time nor training to absorb it in its original form. They are allowed to frame and interpret scientific knowledge and hence represent a substantial source of political power, especially under conditions of scientific uncer-tainty so typical of environmental problems. Empirically, Litfin's study challenges the prevailing view (Benedick 1991; Haas 1992b) that science provided a body of objective and value-free facts from which the ozone regime emerged. Instead, this knowledge was 'brokered' or 'framed' by knowledge brokers. She criticises Haas (1992b) for overestimating the role of information producers and underestimating the role of information framers. Likewise, she emphasises the role of contextual factors more strongly than Haas, particularly the discovery of the Antarctic ozone hole.

Litfin calls for a more interactive and multidimensional conception of knowledge-based power than offered by the epistemic communities literature, proposing an alternative 'discursive practices approach'. She explores the theoretical foundation for such an approach at relative length, focusing on the relationship between knowledge and power (Litfin 1994: 14–51). She argues that post-structuralism, and in particular the later work of Foucault on power and knowledge, provides input for the so-far neglected communicative and generative dimensions of power in the international relations literature on environmental problems. Foucault's argument that power dynamics tend to be more subtle than simple domination or repression is emphasised. Defining the truth through knowledge brokerage equals exercise of power, the argument goes. Discourses, understood by Litfin as sets of linguistic practices and rhetorical strategies embedded in a network of social relations (see p. 6), 'define the range of policy options and operate as resources which empower certain actors and exclude others' (Litfin 1995: 253). Environmental problems are hence 'not simply physical events; they are *discursive phenomena* that can be studied as struggles among contested knowledge claims, which become incorporated into divergent narratives about risk and responsibility' (Litfin 1995: 254).

The role of knowledge brokers is particularly important in so-called 'trans-scientific problems', i.e. problems requiring scientific input to be solved, but that cannot be solved by science alone (Weinberg 1972: 209; Litfin 1994: 30, 204). Studying the achievements of knowledge brokers in trans-scientific problem discourses requires attention to 'the plausibility of their interpretations, the loudness of their voices, and the political context in which they act' (Litfin 1995: 254), or 'the authority of [their] agents, the political context in which [they are] situated, and the cogency of [their] content' (Litfin 1994: 30). Notably, Litfin has a clear conception of active agency,[30] and also places significant emphasis on contextual factors. Methodologically, she calls for the 'intensive case-study approach' of 'detailed contextual analysis' (Litfin 1994: 5–7). She stresses the exploratory nature of this approach, noting that it has little to offer in the way of methodological tidiness and that universally applicable generalisations are not to be expected.

Hajer: story lines and discourse coalitions

Hajer's (1995) point of departure is the claim that most analyses of environmental politics aim to explain how various political interests stand in the way of a real 'ecological turn'. By contrast to this account of fixed actor interests and political outcomes, he claims that:

[I]f examined closely, environmental discourse is fragmented and contradictory. Environmental discourse is an astonishing collection of claims and concerns brought together by a great variety of actors. Yet somehow we distil seemingly coherent problems out of this jamboree of claims and concerns. How does this work? How do problems get defined and what sort of political consequences does this have?

(Hajer 1995: 1–2)

He draws particular attention to how environmental problems are defined through discourse and how the specific structuring of a problem in turn influences the formation of public politics. Hence, although his empirical focus is somewhat broader than Litfin's[31] – aiming in general at the shaping of public politics within a field rather than the formation of a specific regime – they share a preoccupation with how environmental problems are defined or 'framed', and how this affects politics.[32]

In theoretical terms, Hajer (1995: 42–44) refers to the 'communicative miracle' – the entry of social constructivist theory into political science – in introducing his 'argumentative approach'. Drawing on 'social-interactive discourse theory' (Billig 1987; Davies and Harré 1990; Harré 1993), he presents a mild corrective to the work of Foucault. A major premise in this approach is that human interaction is related to 'subject positions' rather than roles. The concept of roles usually implies that a person can always be distinguished from the role he or she assumes. Alternatively, a 'subject position' refers to a person's location in a specific discourse, a position he or she cannot go in and out of as desired. Actors can make sense of the world only by drawing on the terms of the discourses available to them, the argument goes. Hence, persons are *constituted* by discursive practices. Nevertheless, Hajer – like Litfin – has a conception of active agency; persons can actively select, adapt and create thoughts, admittedly within the limits provided by discourses.

Hajer contends that what causes change and permanence is social practices, ascribing both to 'active discursive reproduction or transformation' (Hajer 1995: 56). He takes up Davies and Harré's (1990) concept of *story lines* as a main mechanism in this respect, defining the concept as 'a generative sort of narrative that allows actors to draw upon various discursive categories to give meaning to specific physical or social phenomena' (Hajer 1995: 56). A story line catches certain aspects of a problem complex in a simple and understandable manner. The argument is that people draw on such simplified representations of 'reality' rather than complex systems of knowledge in creating a cognitive comprehension of a subject matter. Story lines play a key role in positioning subjects in a discourse. They re-order people's understanding of problems and hence cause political change. Finding the appropriate story line hence becomes an important mechanism of political agency. Once a story line has been established in a discourse, it settles as 'the way one talks around here'. The disciplinary force of discursive practices often lies in the implicit assumption that others will express themselves within the same discursive frame. Even when their objective is to challenge a dominant story line, people are expected to position themselves in terms of known categories. Society is hence reproduced in a process of interaction between active agents and constraining structures that constantly reinvents the social order. Language is seen not as a medium for subjects to express their preferences, but as a communicative practice which influences actors' perception of interests. Language in use can create new meanings, new subject positions and new politics!

Discourse analysis should, according to Hajer (1995: 60ff), study which cognitive and social traits are routinely reproduced, and the way in which these routinised proceedings are interrupted. Focus should be on those discursive practices that are 'inter-discursive transfer points' where actors change positional statements and new discursive relationships and positionings are created. In addition to story lines, the notion of *discourse coalition* is put forward as a middle-range concept to show how discursive orders are maintained and transformed. A discourse coalition is defined as 'the ensemble of (1) a set of story lines; (2) the actors who utter these story lines; and (3) the practices in which this discursive activity is based' (Hajer 1995: 65). Discourse coalitions differ from traditional political coalitions in several respects: first, their basis is formed by story lines, not interests; second, they broaden the scope for political actors by emphasising the settings where story lines are produced (e.g. the media and science). Hajer's proposed methodology resembles traditional linguistic discourse analysis with its emphasis on the detailed study of texts and interpersonal communication.

Dryzek: the 'grand' discourses on the environment

Dryzek's (1997) aim is to give an overview of major environmental discourses throughout the twentieth century. He is more oriented towards categorisation of the main arguments within the various discourses than towards meticulous case-study description of empirical events and towards embedding his discussion in larger theoretical frameworks. He notes in the introduction to his book that his own work 'will lack the richness of Hajer and Litfin inasmuch as I cannot always say exactly who said what and why behind which closed doors to whom else about a particular point, and how the other responded' (Dryzek 1997: 9).[33] Nevertheless, he shares with the two others a preoccupation with the claim that 'the way we think about the environment can change quite dramatically over time, and [that] this has consequences for the politics and policies that occur in regard to environmental issues' (Dryzek 1997: 5). Hence, questions of change versus continuity in social practices and the relationship between language and politics are at issue here as well.

Dryzek, who rather loosely defines a discourse as 'a shared way of apprehending the world' (see p. 5), contrasts environmental discourse, i.e.

environmentally oriented discourse, with the long-dominant discourse of industrial society, which he calls *industrialism*. Industrialism is characterised by a general commitment to economic growth and increase in the material well-being of a society and its inhabitants. Environmental discourses depart from the terms set by industrialism. This departure can be reformist or radical and it can be prosaic or imaginative, forming two dimensions for categorising environmental discourses. Prosaic departures take the political and economic conditions of industrial society as more or less given. Environmental problems are seen as requiring action, but not in terms of rearranging the entire political-economic chessboard of society. Action can be dramatic and far-reaching, but is confined to the solutions proposed and tools defined by industrialism, mainly through science and administrative bureaucracies. By contrast, imaginative departures seek to redefine the chessboard. Environmental concerns are seen as not necessarily opposed to economic ones, but potentially in harmony. They are perceived as intrinsic to a society's cultural, moral and economic systems and not as difficulties found outside these systems. Like prosaic departures, changes can be either reformist or radical: see Table 1.1.

Environmental problem solving takes the political and economic status quo as given but in need of adjustment to cope with environmental problems. *Survivalism*, a discourse which gained momentum during the 1970s, also sees solutions to environmental problems only in terms of the options of industrialism, but involves a higher degree of change. The 1980s saw the rise of the *sustainability* discourse, which foresees imaginative attempts at solving the conflicts between environmental and economic concerns, but only in a reformist way. Finally, *green radicalism* is both radical and imaginative. Proponents of this perspective discard the basic structures of industrial society and its conception of the environment and propose radically new ideas about the interrelationship between humans and the environment.

Dryzek's approach involves examining the *basic entities, assumptions about natural relationships, agents and their motives* and *key metaphors* of various discourses. He refers to questions about a discourse's basic entities as the ontology of a discourse. Different discourses see different things in the world. Some see nature as brute matter, others conceive of the environment as a self-correcting ecosystem with something similar to intelligence. Some see groups of people (e.g. states) as the basic human entity, others see

Table 1.1 Classifying environmental discourses defined in terms of departures from industrialism

	Reformist	*Radical*
Prosaic	Problem solving	Survivalism
Imaginative	Sustainability	Green radicalism

Source: Dryzek (1997).

individuals or break this category further down on the basis of gender. Furthermore, different discourses embody different notions of what is natural in the relationships between entities, e.g. struggle or co-operation. In the same manner, discourses differ in their conception of agents and their motives. Where some see rational, egoistic humans, others see enlightened individuals with a potential for maximising the future common good by engaging in communicative action and co-operation. Finally, discourses employ different metaphors and other rhetorical devices, e.g. the grazing commons of a medieval village ('the tragedy of the commons'), spaceships ('Spaceship Earth'), human intelligence (ascribed to non-human entities such as ecosystems) or divinity ('Mother Earth').

Like Litfin and Hajer, Dryzek is oriented towards revealing the frames of 'what can be said and done' within a discourse and how this affects actual politics, hence subscribing to most discourse analysis' claim that language in use influences policy choices. Like them, he departs slightly from Foucauldian philosophy in his belief in active agency and his insistence that just because something is socially interpreted does not mean it is unreal.[34] His work is less challenging theoretically and less detailed in empirical terms, but functions as a useful backdrop to more focused studies of environmental discourse.

An agenda for research

The main empirical objective of this book is to elucidate relations between Russia and the West in Arctic environmental affairs during the 1990s. The case studies that are presented relate to specific environmental 'problem complexes' of great importance to the region, namely the management of marine living resources, nuclear safety and industrial pollution. These are 'problems' that affect both Russia and its Western neighbours and are dealt with in a range of bilateral and multilateral environmental regimes. A main goal of the study is to trace similarities and differences between Russian and Western perceptions of these problems, what causes them and how they arc being dealt with. In line with the three approaches presented in the preceding section, focus is on how environmental problems are framed and how this affects politics. All three case studies relate to problems that are dealt with bilaterally or multilaterally in international regimes of which both Russia and one or several Western countries are part. Hence, a main focus is on the interface between Russian and Western (mainly Norwegian) problem definitions, perceived interests and policy choices. How are the (potentially) different perceptions subsumed at the international level and what is the political outcome?

The discussion is organised around discourses on *knowledge, interests* and *institutions*.[35] The assumption underlying this categorisation is, first, that the definition of what should be perceived as 'correct' knowledge is often the basis for policy choices in the management of trans-scientific problems. As emphasised by Litfin, environmental problems are discursive phenomena that can be studied as struggles among contested knowledge claims. Here, we will ask which knowledge claims are produced by relevant scientific circles in Russia and the West and, more importantly, how these are brokered by non-scientific actors in the management process. Second, it is to be assumed that the idea of national interests is somehow intertwined in discussions that involve the distribution of economic benefits and burdens between two or more states. To what extent is it so in the three case studies discussed here? Do national interests loom larger in the debate in Russia or in the West? Are national interests brought forward in the guise of knowledge claims? To what extent do the parties view national interests as underlying the arguments of the opponent? Third, the parties' perceptions of the chosen organisational forms, or institutions, set up at the international level to combat the specific environmental problems are discussed.[36] Do they view the various institutions and practices for trans-national interaction at the bilateral and multilateral level as suitable for these purposes? If not, what kind of objections do they have? Do these objections differ systematically between relevant actors in Russia and the West?

Hence, my main ambition is, in Litfin's words (see p. 9), to explore how discourses 'define the range of policy options' in environmental management. How are the limits defined in the three case studies of what is to be perceived as legitimate knowledge, actor (including national) interests and institutional arrangements? In this discussion, I draw partly on Litfin's conception of knowledge brokerage, partly on Hajer's ideas of story lines and the positioning of subjects, and to some extent also on Dryzek's categorisation of various environmental discourses. Moreover, I have a major focus on the *embeddedness* of environmental discourses in more overarching discourses in society, compare Neumann's (2001) emphasis on this above. It is to be assumed that environmental problems are perceived, talked and written about in familiar categories by subjects both in Russia and the West. I will explore whether environmental discourses in the European Arctic resemble more general discourses in society in Russia and the West and, in particular, in East–West relations in the area.

Methodological considerations

It has already been noted that discourse analysis can provide little in terms of methodological tidiness and universal generalisations. Methodologies proposed include genealogy and 'detailed contextual analysis', or, as Dryzek somewhat mockingly refers to Litfin and Hajer, accounts of 'who said what to whom behind which closed doors and how the other responded' (see p. 11). In addition to written material such as official documents and press reports, I will draw on my personal experience from nearly one and a half decade's preoccupation with environmental affairs in the European Arctic as a student, researcher, interpreter, inhabitant in the Western part of the region and frequent visitor to its Russian part. As a student and a researcher, I have conducted a range of investigations on the management of marine living resources, nuclear safety and more general environmental problems in the region.[37] These studies have entailed hundreds of interviews with relevant actors in both Russia and the West as well as a fair amount of observation of ongoing practices in the management of the environment.[38] As a Russian-language interpreter, I have been a first-hand witness to much of the interaction taking place between Russians and Scandinavians in the environmental sphere, particularly in the management of marine living resources.[39] While living in Northern Norway in the early 1990s and as a frequent visitor to Northwestern Russia throughout the 1990s, I have come close to the environmental problems themselves and to people who are directly affected by them.[40]

My method is an 'attempt at genealogy', realising that this approach ideally requires a more thorough treatment of each of my three case studies than will be provided here. It is nevertheless my ambition to trace 'nooks and crannies', drawing on all available material. I usually start out with written texts to investigate discourses on knowledge related to the environmental problem in question. I then move on to various types of official documents to inquire whether the different bits of scientific advice are reflected in official statements by the authorities in Russia and the West, and in protocols and minutes from meetings at the international level. Views on national interests and institutional arrangements are primarily reported from the press, formal interviews and informal conversations that I have conducted with relevant actors.

I have elsewhere elaborated on the difficulties of conducting social science research in Northwestern Russia (Hønneland and Jørgensen 1999a: 6–7): being one of the most heavily militarised areas in the world and rather 'conservative' politically,[41] the region is in general characterised by a suspicious attitude to curious foreigners. Since Russia has no long-standing tradition of 'free' social science research, foreign researchers are often mistaken for journalists or even spies. Furthermore, foreign involvement in environmental affairs in the region has, since the arrest of Aleksandr Nikitin in February 1996,[42] come to be closely associated with espionage in the minds of many inhabitants of Northwestern Russia.[43] In addition to a careful and gradual cultivation of trust among potential sources of information, the well-known Russian method of using acquaintances has been indispensable for this study.[44]

As a result of this suspicion by certain Russian authorities, I have found it necessary to anonymise my interviewees. Instead of referring to them by name, I indicate their professional background, e.g. 'representative of Russian regional authorities', 'Norwegian NGO activist' or 'Russian scientist'. This is done mainly since a precondition for many of the interviews was that the names of the interviewees should not be revealed. Also where such an agreement was not explicit, I have chosen anonymisation in order to protect my Russian interviewees from potential problems at home. Occasionally, I also refer to conversations of a more private character, to utterances I have incidentally come across, and to events I have observed. In some of these cases, I have made slight changes in the context of the conversation or observation – again to prevent those involved from being recognised. Some might find this methodologically dubious; however, it is not uncommon in the anthropological literature, e.g. in describing social practices in a given society without identifying either the society or its inhabitants. If minor changes are made in the situation in which a statement was made, the substance of the argument remains, of course, unchanged. Some interview statements are referred literally, as noted down by me during the interview.[45] Others were written down only after the interview; hence, the meaning of the statement is, I hope, maintained while these statements should not be regarded as word-for-word representations of what was said.[46] These extracts are indicated in the text as 'conversations' rather than 'interviews'.

The study mainly refers to environmental discourses during the 1990s, with an emphasis on the second half of the decade. The industrial pollution and nuclear safety problems became an issue only in the late 1980s and early 1990s. The management of marine living resources in the Barents Sea has been a bilateral matter between Russia (or the former Soviet Union) and Norway since the late 1970s, but I will limit my discussion to the 1990s here as well.

Outline of the book

Chapter 2 provides some background information to the main topic of the book. It presents the environmental problems discussed in the book and the institutional arrangements set up to combat them in somewhat more detail than in this introductory chapter. Then follow the three case studies on the management of marine living resources, nuclear safety and industrial pollution. Each case study covers discourses on knowledge, interests and institutions. Chapter 6 sums up the main findings of the case studies and resumes the theoretical discussion on environmental discourse.

2 The environment and institutions in the European Arctic

Introduction

The purpose of this chapter is to give a general outline of the environmental problems discussed in the book and the institutional arrangements set up to combat them. An obvious dilemma in providing such an overview is that defining these problems and their solutions lies at the heart of the subsequent discussion. So how can they be defined 'in advance'? And how would doing so influence their discussion in the case studies? I have chosen to provide a review of the environmental status quo as presented in publications of international scientific bodies and in the scientific literature. An overview of the institutional arrangements designed to deal with the problems is given, with reference to how the organisation, activities and performance of these arrangements are generally presented in the scientific literature as well as in publications issued by the organisations themselves. Although an attempt is made to give the outline a factual basis, some amount of 'framing' takes place already at this stage since some features of the environment and institutions are emphasised while others are left out. This is partly a consequence of how these phenomena are presented in the literature and by the organisations themselves and partly active selection on my part. However, it should be kept in mind that this chapter is meant as an introduction to problems that will be further discussed in the case studies. It should therefore be regarded as a point of departure, a preliminary outline of commonly held (admittedly scientific) beliefs that will to some extent be challenged later in the book. A major objective of the subsequent discussion is to investigate how these beliefs have been arrived at and whether they are shared equally in Russia and the West and among various types of actors in the different countries.

This chapter is organised in line with the presentation of case studies, focusing on the management of marine living resources in the Barents Sea, nuclear safety issues in Northwestern Russia and industrial pollution in the European Arctic. Within each of these three fields, an overview is first given of the state of the natural environment, then of the institutional arrangements designed to manage it.

The marine living resources of the Barents Sea

The Barents Sea comprises those parts of the Arctic Ocean lying between the North Cape on the Norwegian mainland, the South Cape of Spitzbergen Island in the Svalbard archipelago, and the Russian archipelagos Novaya Zemlya and Franz Josef Land. Traditionally, the fish and marine mammals of the Barents Sea have provided the basis for settlement along its shores, particularly in Northern Norway and in the Arkhangelsk region of Russia. Since the Russian Revolution in 1917, the city of Murmansk on the Kola Peninsula has functioned as the nerve centre of the Russian 'northern fishery basin', second in importance in the country to its 'far eastern fishery basin'.

Since 1976, the main fish stocks of the Barents Sea have been managed bilaterally by Norway and the Soviet Union/Russian Federation through their Joint Fisheries Commission. The Russian–Norwegian fisheries regime also handles the management of seals in the area while whales are the responsibility of the International Whaling Commission (IWC). The conservation of biodiversity in the Barents Sea has since 1997 been taken care of by the Joint Russian–Norwegian Environmental Commission.

The resource base

The Barents Sea contains a large abundance of fish stocks of a variety of species (see Figure 2.1). The reason for this abundance is the rich plankton production in these waters, providing food for large stocks of *pelagic* fish, i.e. fish living in the space between the seabed and the surface. The pelagic fish stocks, first and foremost capelin and herring, are in turn the prey of *groundfish*, such as cod, haddock and saithe. Both pelagic fish and groundfish serve as food for sea birds, marine mammals and people. Cod, capelin and herring are key species in the ecosystem. Cod feed on capelin, herring and smaller cod, while herring feed on capelin larvae. Periods of growth in the cod and herring stocks and a reduced capelin stock tend to alternate with periods without herring in the Barents Sea, while there is a moderate growth in the cod stock and capelin is abundant.

The Barents Sea capelin (*Mallotus villosus*) stock used to be among the largest and most important fish stocks in the Northeast Atlantic. One major trait of this stock is that large variations in individual growth from year to year lead to substantial fluctuations in stock size, with considerable implications for the whole ecosystem. Since the mid-1970s, the total biomass has fluctuated from close to 9 million tonnes (1975) to 0.2 million tonnes (1986). The size of the capelin stock was rather stable during the 1970s while in the 1980s it fell dramatically. Overfishing reinforced a natural downward fluctuation, bringing the stock close to total breakdown. Commercial fishing for capelin was halted in 1986, and the stock started to recover. Fishing was resumed again in 1991. A new collapse occurred in 1993. The stock has recovered once again – at least partly. Fishing was resumed in 1999 for the first time since 1993. The quota that has been established is quite moderate.

Figure 2.1 The Barents Sea.

The stock of Atlanto-Scandinavian herring (*Clupea harengus*) was the largest fish stock in European waters until its collapse in the late 1960s. The normal migration pattern from the North Sea to the spawning grounds off the Norwegian coast was broken in 1970. The stock had then become so reduced that the remaining fish found sufficient food off the Norwegian coast. The old migration pattern resumed only in the mid-1990s. The stock has increased considerably in recent years, and herring catches in the North Sea are approaching those of the 1950s. However, the stock is now thought to be in decline again. The cohort groups from 1993 onward seem to be weak. Some herring fry drift with the Gulf Stream into the Barents Sea each year, although their numbers are highly variable. Each age group spends three years in the area, feeding on capelin while they drift northward. While contributing to the total herring stock, the young herring of the Barents Sea may be considered a threat to the capelin. However, the herring also relieves the pressure on capelin, since it is itself preyed upon by cod. Although herring is not commercially exploited in the Barents Sea, its presence does affect the fishing in the area.

The Northeast Arctic cod (*Gadus morhua*) spawns along the coastline of Norway from the age of 7. After spawning, they return to the Barents Sea. Fry of this species also drift into the northern parts of the Barents Sea. From the age of 4, cod prey upon capelin as the latter species moves southwards to its breeding grounds. The cod stock reached a total of some 4 million tonnes at the end of the 1960s. The stock decreased steadily throughout the 1970s, ending up at less than 1 million tonnes in 1984. Strong age groups were recruited during the years 1983–1985, but these cohorts fell back in 1986–1988 due to a lack of food and extensive fishing. The level of the late 1970s was not regained until the beginning of the 1990s. This increase was mainly realised through quota regulation, which significantly moderated the fishing effort for cod. The stock of cod has shown a decline since 1993. Furthermore, stock estimates were reduced in 1998, when scientists discovered that their estimation method had resulted in overestimations of the stock. The cod quota has been cut significantly in recent years.

The Northeast Arctic haddock (*Melanogrammus aeglefinus*) stock in the Barents Sea tends to follow the same pattern as the cod. Having reached an all-time low in 1983–1984, an increase was brought about by strong age groups from the period 1982–1983. Another decline took place in the late 1980s. From 1990, recruitment improved markedly until 1995. In recent years, a modest overall decrease has been observed, but the spawning stock has increased. Fluctuations in the size of each age group are more significant for haddock than for cod, and the total stock is also considerably smaller. Thus, it is generally considered to be difficult to sustain a stable haddock fishery over time.

The saithe (*Pollachius virens*) stock has remained at a low level since the mid-1970s. The seal invasions of 1987–1988 generated a crisis for this stock. Since then, a gradual build-up took place, but today, the stock is once again decreasing. However, age cohorts from the early 1990s are quite good. Like haddock, saithe is mainly caught as a by-catch in the cod fishery. The estimations of the redfish (*Sebastes mentella* and *Sebastes marinus*) stocks are rather uncertain, and fishery biologists have refrained from presenting specific proposals regarding quotas. There was an intensive redfish fishery up to 1991. A targeted fishing effort for this stock took place southwest of Bear Island in the spring. This probably led to a significant decline in stock size. While general estimates are vague, it has been established that recruitment has been alarmingly low since 1991. Fishing for redfish has fallen sharply in recent years due to small catches. It is generally assumed that it will take considerable time for the stocks to recover. Redfish grows slowly and its fry are preyed upon by cod and herring. The direct fishing for Greenland halibut (*Reinhardtius hippoglossoides*) was stopped in 1992 because the stock was near extinction. It is now taken primarily as a by-catch. A limited amount of inshore fishing using passive gear is permitted. The strict regulations introduced in 1992 resulted in some recovery of the stock, but this trend was reversed when several weak age cohorts became sexually mature. Direct fishing for halibut is not expected to be resumed in the foreseeable future.

The large abundance and diversity of marine mammal species in the Barents Sea attracted the attention of the earliest European explorers of the region. Massive harvests, beginning in the seventeenth century, targeted various species of seals and whales, at different times, over a period of more than 300 years. These various periods of exploitation resulted in the decimation of many populations of marine mammals in this part of the Arctic, and the extirpation of some species from some areas. For example, the walrus was brought to the verge of extinction on Svalbard during the mid-1950s after a period of exploitation beginning in the early 1800s, and bowhead whales were practically driven to extinction both within this region and throughout the rest of the Arctic. Currently, only harp and hooded seals and minke whales are subjected to significant commercial harvests within the Barents Sea region. However, all coastal seals are hunted in both Norway and the western Russian Arctic, and white whales are still harvested to some degree in the Kara Sea. Small local catches are also known to occur for small toothed whales along the mainland coasts, although harvests are small and no statistics are available.

The Russian–Norwegian fisheries management regime

The United Nations Conference on the Law of the Sea (UNCLOS) (1975–1982) led to a transition from multilateral negotiations for the Barents Sea fisheries under the auspices of the Northeast Atlantic Fisheries Commission (NEAFC) to bilateral negotiations between coastal states with sovereign rights to fish stocks. Norway and the Soviet Union entered into several bilateral fishery co-operation agreements in the mid-1970s. The Russian–Norwegian management regime for the Barents Sea fish stocks defines objectives and practices for co-operative management between the two states – in addition to national-level management procedures – within the fields of research, regulations and compliance control. The co-operation between Russian/Soviet and Norwegian scientists in the mapping of the Barents Sea fish resources dates back to the 1950s. It is now institutionalised within the framework of the International Council for the Exploration of the Sea (ICES). The main participants are the Knipovich Scientific Polar Institute for Marine Fisheries and Oceanography (PINRO) in Murmansk, the Norwegian Institute of Marine Research in Bergen and the Norwegian Institute of Fisheries and Aquaculture Ltd in Tromsø. The institute in Tromsø has in recent years undertaken the main responsibility for research on shrimp and marine mammals, while the other species of the Barents Sea are largely the responsibility of the Bergen institute. The co-operation effort is generally characterised as successful (Stokke *et al.* 1999).

The Joint Russian–Norwegian Fisheries Commission includes members of the two countries' fishery and other authorities, marine scientists and representatives of fishers' organisations. The Norwegian delegation to the commission is headed by the administrative leader of the Norwegian Ministry of Fisheries. The Russian delegation is headed by the first Deputy Chairman of the Fisheries Committee of the Russian Federation. The Commission meets at least once a year, establishes total allowable catches (TAC) for the joint fish stocks of the Barents Sea: cod, haddock and capelin. Cod and haddock are shared on a 50–50 basis, while the capelin quota is shared 60–40 in Norway's favour. In addition, quotas of the parties' exclusive stocks are exchanged. Russia has traditionally given a share of its cod quota to Norway in return for a share in Norway's quotas of redfish, herring and Greenland halibut. However, after the introduction of reforms in the Soviet Union in the late 1980s, the Russians have kept a larger portion of their cod quota. After the sessions in the joint commission, the two parties conduct further quota exchanges in bilateral negotiations with third countries. Traditionally, the Soviet Union/Russian Federation has given part of its Barents Sea cod quota to the Faeroes, while Norway has transferred a share of its quota to the EU in exchange for quota shares in the North Sea.[1]

The Norwegian Coast Guard exposed a dramatic increase in under-reporting by Russian vessels in 1992, and took extra steps to calculate the total Russian catch in the Barents Sea for that year. By the end of 1992, Norwegian fishery authorities had presented these figures to their Russian colleagues. They indicated Russian overfishing of more than 100,000 tonnes. Overfishing constituted one-quarter of the total cod quota in the Barents Sea in 1992. Russia had 170,000 tonnes at its disposal, of a TAC of 396,000 tonnes, after internal quota exchanges with Norway. This estimate was supported by export statistics, which indicated that nearly the entire Russian cod quota in the Barents Sea had been exported to Norway. At the same time, considerable quantities had been exported to other Western countries. Some cod had also been landed in Murmansk. This sudden rise in overfishing coincided with Russian fishers starting to deliver the bulk of their catches abroad, primarily in Northern Norway. This direct export of their product also increased the incentives for fishers to underreport their catches, and reduced the ability of Russian authorities to keep track of the catches since control had traditionally been conducted in connection with landings of fish in Russian ports.

Towards the end of 1992, both Norwegian and Russian authorities had become aware of the shortcomings of the control of Russian fisheries in the Barents Sea. At the twenty-first session of the Joint Russian–Norwegian Fisheries Commission in November 1993, the delegation leaders of the two parties jointly proposed the appointment of an expert group to consider the question of co-operation between the control bodies of the two states. The expert group presented its recommendations in May 1993, and the recommended measures were soon implemented. The most important initiative was to establish a formal exchange of catch information. The Norwegian Directorate of Fisheries now regularly sends data of all Russian landings in Norway to control authorities in Murmansk. Additionally, inspection data may be submitted on request. An organised exchange of information about activities at sea has been established between the Russian control body Murmanrybvod and the Norwegian Coast Guard. Moreover, an exchange of inspectors is regularly carried out; most importantly, Russian inspectors have been allowed to participate as observers when inspectors from the Norwegian Fish Control (a branch of the Directorate of Fisheries, responsible for catch control ashore) inspect Russian vessels in Norwegian ports. Upon delivering its recommendations, the expert group was transformed into a permanent working group for Russian–Norwegian co-operation on control and management issues under the Joint Fisheries Commission, normally referred to as the Permanent Committee. It has met at least once a year since 1993. It is headed by representatives of the Directorate of Fisheries (Norway) and the regional control body Murmanrybvod (Russia). It also includes representatives of the Norwegian Coast Guard and the old fishery complex association Sevryba and PINRO in Russia.[2]

During the 1990s, Russian–Norwegian fisheries management co-operation was generally assessed to be fairly successful (Hoel 1994; Stokke 1995; Hønneland 1998a). Most stocks had grown steadily since the end of the 1980s. The capelin stock, with its rapid fluctuations, represented a departure from this tendency. While the stock had quickly recovered from almost total breakdown in 1985–1986, a new setback loomed in 1992–1993. Furthermore, the Greenland halibut stock suffered a threatening decline since the early 1990s, and is still taken only as by-catch (except for a limited direct fishery with passive gear). The TAC of Northeast Arctic cod amounted to 700,000–850,000 tonnes in the mid-1990s. A decline in the stock has been observed since then (see p. 20). The quota was accordingly reduced from 850,000 tonnes in 1997 to 480,000 tonnes in 1999. It should be observed that even the latter figure was far above the scientific recommendations of ICES for that year. Other problems in the co-operation process between Norwegian and Russian authorities in recent years include the repeated refusals by Russian authorities to requests for joint Russian–Norwegian research cruises in the Russian Zone of the Barents Sea. However, it should be noted that these refusals come from Russian security services, and not from fishery management authorities. On the positive side, an extension of the regime to include co-operation between control bodies from the two states has been made, and Norway and Russia took a common stance against Iceland in the dispute regarding the Barents Sea Loophole, an area of international water outside the 200-mile zones of Norway and Russia.

Nuclear safety in the European Arctic

As we saw in Chapter 1, a major environmental threat to the European Arctic in recent years has been the danger of radiation from nuclear installations, discarded nuclear vessels, radioactive waste and spent nuclear fuel in Northwestern Russia. The build-up of the 'nuclear complex' in North-western Russia was linked partly with the nuclearisation of the Soviet Armed Forces, partly with civilian needs in transportation and energy production. Its history dates back to the late 1950s, when the Soviet Union's first nuclear-powered submarines and icebreakers were constructed and stationed on the Kola Peninsula. Well over a decade later, in 1972, the construction of the largest nuclear installation in the area, the Kola nuclear power plant, was completed. Thus, the nuclear-powered vessels of the Northern Fleet, the ice-breaker fleet of Murmansk Shipping Company and the Kola nuclear power plant, together with their service and storage facilities, constitute the main elements of the Northwest Russian nuclear complex. A brief description of the development of each of these elements and their conditions today is given in the first sub-section.

Perhaps contrary to popular belief, the extent of radioactive contamin-ation in the region is relatively low and mainly due to external sources and practices that have been permanently or temporarily discontinued, i.e. atmospheric nuclear tests and the dumping at sea of radioactive waste (Lønne *et al.* 1997; Bergman and Baklanov 1998). The issue of contamin-ation and its sources is elaborated in the second sub-section. While the release of radioactive substances is negligible in the normal operations of the nuclear installations in Northwestern Russia, an accident or other unforeseen incident at one of the installations, service or storage facilities could have very serious implications. The risks posed by the nuclear complex are addressed in the third sub-section. Finally, a brief overview is given of the international efforts made during the 1990s to contribute to nuclear safety in the European Arctic.

The 'nuclear complex' of Northwestern Russia

The Soviet Northern Fleet was officially established as late as in 1933, despite the fact that the strategic potential of the Kola Peninsula, with its year-round ice-free harbours, had been recognised in Russian naval circles as early as in the second half of the nineteenth century (Skogan 1986). The fleet headquarters were originally based in Polyarnyy on the western shore of the Kola Fjord, but subsequently moved to Severomorsk across the fjord. For more than two decades, the Northern Fleet remained the smallest of the Soviet Navy's four fleets. In the 1950s, however, a period of expansion set in, which coincided with the onset of the Soviet Union's struggle to achieve nuclear parity with the United States. The country's first nuclear-powered submarine was completed at a submarine construction facility in Severodvinsk in Arkhangelsk Oblast in 1957 and stationed at Zapadnaya Litsa on the Kola Peninsula in 1958. The Northern Fleet's expansion continued in the following decades, and the fleet acquired a large number of nuclear-powered submarines.

The fleet's prize possessions are its strategic submarines. They are armed with long-range ballistic missiles able to reach targets on the American continent from the Russian coast. The strategic submarines are protected by multi-purpose submarines, which are also nuclear-powered. The tactical missiles of these submarines can be equipped either with nuclear or with conventional warheads. Russia has declared that the multi-purpose submarines will not carry nuclear warheads in peacetime. In addition to the submarines, the Northern Fleet has a large number of surface vessels, many of which are armed with nuclear weapons and three of which are nuclear-powered.

The Northern Fleet reached the height of its power in the mid-1980s. Since 1988, the number of ships taken out of service has exceeded the number of new ships acquired, a tendency that intensified in 1991. This was due to the combined effects of a period of economic downturn and the need to implement obligations laid down by the START-1 treaty. In addition, the nuclear submarines of the first generation started to reach the end of their service life from the late 1980s. A total of about seventy nuclear-powered submarines were removed from service in the Northern Fleet during the period 1991–1998 (see Table 2.1). This gave the Northern Fleet a new challenge: to dismantle a huge number of nuclear-powered submarines in a very short period of time. The process is very costly and, so far, only a fraction of the submarines removed from service have actually been dismantled. The majority are stored afloat awaiting dismantlement, and many have not even been defuelled, i.e. the spent nuclear fuel has not been removed from the reactors. Some of these submarines are in a poor condition, and some observers claim there is a real danger of them sinking. Moreover, international support programmes so far cover only the dismantlement of strategic submarines. Hence, there is no dismantlement capacity left for multi-purpose submarines, which are mostly at the end of or have passed far beyond their service life.[3]

The full dismantlement of a nuclear-powered submarine involves several

Table 2.1 Number of ships in the Northern Fleet in 1991 and 1998

Category	1991	1998
Strategic submarines	36	13
Other submarines	90	43
Large combatants	83	45
Smaller combatants	73	27
Landing and amphibious vessels	20	12

Source: Hønneland and Jørgensen (1999a).

steps. The first step – and the most important from an environmental point of view – is the defuelling of spent nuclear fuel from the submarine's reactor. Afterwards, the reactor compartment is removed, and subsequently the rest of the hull can be cut up for scrap. The process generates a certain amount of spent nuclear fuel and other liquid and solid waste (including the reactor compartment itself) which has to be transported to special facilities for further treatment and/or intermediate storage. Spent nuclear fuel can be either stored as it is or reprocessed. The Russian policy with regard to spent nuclear fuel favours reprocessing, if possible. Reprocessing generates large amounts of high-level liquid waste. Liquid waste is usually solidified before storage, and some forms of solid waste can be compressed.

The Northern Fleet has limited capacity in all stages of the dismantlement process. First, the existing infrastructure has allowed for the defuelling of only a handful of submarines a year. Second, existing intermediate storage facilities for spent nuclear fuel and for liquid and solid waste are long since filled to capacity, which means that large amounts of spent nuclear fuel and radioactive waste are stored under unsatisfactory conditions. Third, the spent nuclear fuel that is supposed to be reprocessed is leaving the region very slowly, partly because Russia has only two special train sets for spent nuclear fuel transportation, partly due to constraints at the reprocessing facility Mayak in Siberia and to the Northern Fleet's limited capacity to pay for such transportation.

Up to 1992, although the situation was less critical then, the Northern Fleet disposed of large amounts of radioactive waste by dumping it at sea. Since 1993, Russia has observed a voluntary moratorium on such dumping, but the lack of capacity to deal with spent nuclear fuel and nuclear waste has, according to Russian authorities, rendered it impossible for Russia to ratify the parts of the London Convention that deal with the dumping of radioactive substances. The problems related to the decommissioning of nuclear-powered submarines is the most urgent nuclear safety issue in the military sector. However, as demonstrated by the *Kursk* and *Komsomolets* accidents, there are risks associated with nuclear-powered submarines in operation as well. Moreover, nuclear-powered submarines normally have to have their reactors refuelled twice in the course of their service life; thus, active submarines also contribute to the accumulation of spent nuclear fuel and to binding up defuelling infrastructure. These issues will be discussed further towards the end of the section.

In 1959, the Soviet Union's first nuclear-powered icebreaker – the *Lenin* – was completed and stationed in Murmansk. Over the next few years, a whole fleet of nuclear icebreakers was built.[4] The main task of the icebreakers was to escort vessels navigating the Northern Sea Route (i.e. the North-East Passage), which stretches along the northern shores of the Eurasian continent from the Barents to the Bering Sea. Thus, the icebreaker fleet played an important role – and to some extent still does – in securing the *severnyy zavoz*, i.e. the transportation of foodstuffs and other important goods to outlying northern areas, as well as the transportation of various raw materials out of these areas, mainly from Siberia.

The icebreaker fleet is a subsidiary of Murmansk Shipping Company, one of the largest enterprises in Murmansk Oblast. The icebreaker fleet, which in 1998 counted seven icebreakers and one nuclear-powered container ship, was needed less and less during the 1990s, because transport volumes on the Northern Sea Route fell significantly. The remaining ships are approaching the end of their service life. Like the nuclear-powered submarines, the icebreakers have to have spent nuclear fuel removed and reactors refuelled at regular intervals. Murmansk Shipping Company has three service ships of its own, *Imandra*, *Lotta* and *Lepse*. Spent nuclear fuel from the icebreakers used to be stored for an initial six months on board the *Imandra* and then moved to one of the other ships for further storage. However, by 1993, all three ships were filled to capacity. As for *Lepse*, the whole ship is now considered nuclear waste, since a large portion of the spent nuclear fuel on board is classified as 'damaged', and the ship itself is also contaminated. This damaged fuel stems mainly from the nuclear-powered icebreaker *Lenin*, which suffered a reactor accident in 1966.

The Kola nuclear power plant, located in the town of Polyarnye Zori in the southern central part of the Kola Peninsula, was put into operation in 1972. The plant is by far the largest employer in Polyarnye Zori with approximately 6,000 employees (out of a population of some 18,000). As was customary in the Soviet Union, the company provided a large share of the local social and technical infrastructure, and many of these functions have been retained. The Kola nuclear power plant and the eight other nuclear power plants currently in operation in Russia are joined together under the Rosenergoatom umbrella. Rosenergoatom is a subsidiary of Minatom, the Russian Ministry of Atomic Energy, which reportedly wants to see closer integration of the enterprises of the Russian nuclear energy sector. In June 1999, then Minister of Atomic Energy Yevgeniy Adamov decided to let Rosenergoatom have direct control of the revenue generated by all nuclear power plants. Plans to continue the reorganisation of the nuclear power sector towards full integration have met with considerable opposition at the regional and local levels. Workers at the Kola nuclear power plant fear that both their pay checks and general industrial rights will suffer, while local as well as regional authorities expect the merger process to result in significant cuts in revenue from the Kola nuclear power plant to the local and regional coffers.

The Kola nuclear power plant plays a significant role in the total supply of energy in Murmansk Oblast (see Figure 2.2). Total yearly output is approximately 12 terawatt hours (TWh). Altogether, the Kola nuclear power plant provides some 60 per cent of the electric power consumed annually in the oblast (Hansen and Tønnessen 1998). Most of this is consumed by the industry, since electric power accounts for only a marginal share of the energy consumed by private households. In addition, the Kola nuclear power

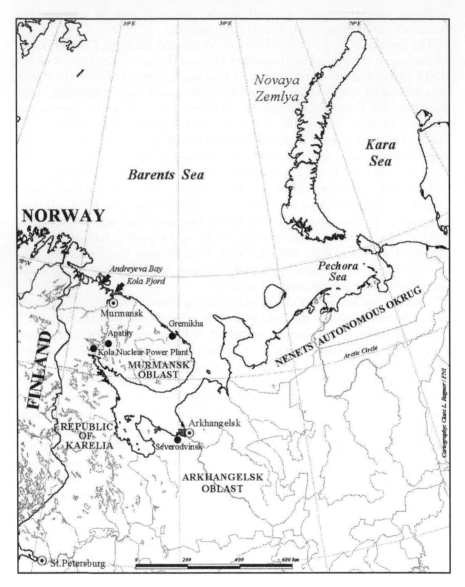

Figure 2.2 Important Northwest Russian sites in a nuclear safety context.

plant exports energy to the Republic of Karelia. The plant has four nuclear reactors, each of which yields an effect of 440 megawatts (MW). The two oldest ones, contained in one building, came on line in 1973 and 1974, respectively, while the two reactors contained in the other building have been in operation since 1981 and 1984. The two oldest reactors in particular are viewed with concern by the international community.

Radioactive contamination of the environment

The level of radioactive contamination of the terrestrial environment in Northwestern Russia is relatively low and comparable to that of neighbouring countries (AMAP 1997, 1998; Bergman and Baklanov 1998; Lønne *et al.* 1997). As in Norway, Sweden and Finland, the main sources of contamination are fallout from atmospheric nuclear tests and the Chernobyl accident. A total of eighty-seven atmospheric tests were carried out on the Novaya Zemlya archipelago between 1955 and 1963 (Stortinget 1994: 31). Since 1963, subsequent to the conclusion of the partial test ban treaty prohibiting atmospheric and underwater nuclear tests, all nuclear tests have been carried out underground.[5]

Despite the low contribution to the average level of contamination by regional sources, the localised character of such contamination must be taken into consideration. While fallout both from the nuclear tests and Chernobyl was dispersed over vast areas, more recent activities – particularly inadequate storage of waste – have resulted in a number of delimited areas where contamination levels are significant. On the Kola Peninsula as a whole, though, the main anthropogenic source of radiation affecting the population is radiotherapy, followed by radiation from construction materials. By contrast, production of nuclear energy and fallout from weapons testing and the Chernobyl accident make up less than 3 per cent of the average yearly individual doses to the population (State Committee for Environmental Protection 1998).

Compared to the Baltic Sea and the North Sea, the Barents and Kara Seas are very clean. This goes for radioactive contamination as well as for pollution in general. Besides fallout from atmospheric nuclear tests, the main sources of radioactive substances in these waters are routine releases from European reprocessing plants, first and foremost Sellafield in Great Britain. In recent years, contamination levels have been reduced, primarily due to a sharp reduction in releases from Sellafield. Releases from Russian facilities are also believed to have contributed to contamination increases via the Ob and Yenisey rivers. The total radioactivity of materials dumped by the Northern Fleet (until 1992) and Atomflot (until 1986) is comparable to the accumulated radioactivity of the Sellafield releases. However, all investigations carried out thus far have concluded that very little of this has actually leaked out. Moreover, even in the event of a worst-case scenario, with a simultaneous and rapid release of the remaining radioactivity to the environment, radiological consequences would be relatively low (Bergman 1997).

Radiological hazards associated with the nuclear complex

The principal radiological hazards to the region as well as to areas beyond it are posed not by previous activities but by potential future accidents or incidents. The risks posed by a release of radioactive substances to the life and health of humans fall into two main categories, depending on the distance from the site where the accident or incident takes place. Persons at or near the site will likely be in acute danger from intense radiation, as well as from other effects of the accident, such as fires and explosions. At greater distances from the site, the main dangers are usually related to long-range effects of more moderate exposure, for instance increased frequencies of certain forms of cancer in the exposed population.

It is problematic to distinguish between environmental risks and health risks where radioactive pollution is concerned. For a nuclear accident to constitute a real threat to the environment as such – that is, to the continued existence of various species of flora and fauna in a given area – the accident would have to be of almost unimaginable proportions. Usually, the main problem is that the presence of radioactive pollutants in the natural environment may pose a risk to people's life and health. The most seriously affected areas may be rendered uninhabitable. Moreover, as demonstrated by the Chernobyl accident, the problem of radioactive pollutants entering and accumulating in the food chain may affect vast areas, making certain foodstuffs – typically the meat of grazing animals – unfit for human consumption. Thus, in addition to the risks to life and health, radioactive contamination of the environment may have grave economic consequences, and it may even threaten the culture and way of life of certain people, notably those of indigenous peoples and fishers.

A study by Bergman and Baklanov (1998) substantiates the Kola nuclear power plant's position as the most hazardous industry in the Kola-Barents region. Several operational incidents have taken place since the plant was put into operation, among them a loss-of-cooling incident in 1993, which might have resulted in a meltdown incident in the oldest reactor. The International Atomic Energy Agency (IAEA) has calculated the probability of a serious meltdown in the oldest reactors to be 25 per cent over a period of 23 years. Such an accident would not have Chernobyl dimensions, since the reactors are of a different and less dangerous design. Nevertheless, the local effects would be grave, and deposits of radioactive matter in parts of Russia and neighbouring countries would probably be high enough to influence both cancer statistics and patterns of food consumption.

Bergman and Baklanov (1998) identify submarines during refuelling as another high-risk object, although the consequences of an accident would be less far-reaching. In 1985, radionucleids were released to the atmosphere and ten people were killed following an explosion during refuelling of a submarine in Chazma Bay in the Far East. For submarines in operation, under decommissioning or scrapping, and for stored spent nuclear fuel and radioactive waste, the risks are less certain. These are all classified by Bergman and Baklanov (1998) as sources of potential high risk.[6] Where the submarines are concerned, the main dangers are linked to criticality accidents (i.e. uncontrolled chain reactions) in reactors with releases of radionucleids to air. Defuelling of decommissioned submarines may involve the same risks as refuelling. Accidents involving inadequately stored spent nuclear fuel and radioactive waste may ensue as a result of fire breaking out at a storage facility and/or a criticality accident if spent nuclear fuel elements are stored in too close proximity to one another and without sufficient barriers between them.

Spent nuclear fuel and other waste dumped at sea, as well as the sunken submarine *Komsomolets* and leakage from underground test sites, are all categorised as low-risk sources by Bergman and Baklanov (1998). Finally, their study identifies some potential risks where the consequences are as yet undetermined due to insufficient research. They include transportation of spent nuclear fuel and waste, accidental nuclear explosions and releases from underwater test sites.

International nuclear safety efforts

The real or imagined threat of radiation from Northwestern Russia has caused a lot of public concern in Northern Europe during the 1990s, especially in the Nordic countries. It has also led to the establishment of several international co-operation arrangements aimed at reducing the threat at both the bilateral and multilateral level. A Norwegian Plan of Action for Nuclear Safety in areas adjacent to Norway's borders has been in effect since 1995 (Ministry of Foreign Affairs 1995). Its overriding goals are to protect health, the environment and business against radioactive contamination and pollution from chemical weapons. The activities under the Plan of Action are categorised into four prioritised areas:

- safety measures at nuclear facilities
- management, storage and disposal of radioactive waste and spent nuclear fuel
- radioactive pollution in northern areas
- arms-related environmental hazards.

As of January 2000, some 113 projects were listed under the Plan of Action with total budgets of some 536 million NOK. During the period 1995–1999, only 343 million NOK was actually spent on the Plan, since many of the projects were not yet started or were delayed. Priority was given to co-operation on safety issues with the Kola nuclear power plant, investigations and evaluation of pollution in northern areas and reviews of the cost-effectiveness of measures to alleviate the situation in certain key areas. A Framework Agreement was signed between Norway and Russia and a Joint Commission on Nuclear Safety to oversee its implementation set up in 1998.

Some 68 per cent of the total funds allotted to the Plan of Action during the period 1995–2000 were given to construction of facilities, but a considerable part of this money was not spent. The overwhelming share of funds was used or planned to be used in Russia. Still, a sizeable sum was given to activities in

Norway, for instance for research and competence building in the field of nuclear safety. Concrete project proposals come from the various mechanisms that have been established for bilateral and multilateral co-operation with Russia. The development of projects in isolation from each other is a serious challenge, in particular with regard to the physical handling of nuclear material. Progress in one link in the chain may be futile if the next link is missing. The lack of co-ordination between various authorities in Russia exacerbates this problem (Hønneland and Moe 2000).

It has been an explicit Norwegian goal to play a catalytic role in raising international awareness of and financial support to nuclear safety in Russia, and to create proper multinational mechanisms for this purpose. One major venture in this respect, initiated by Norway, is the trilateral Arctic Military Environmental Co-operation (AMEC) between Norway, Russia and the USA. This co-operation arrangement, which has been in place since 1996, is directed towards military-related environmental issues in the Arctic. The parties of the agreement have stated their mutual interest in reducing the deleterious effects of military operations to the Arctic environment, including the ecological risks associated with nuclear waste in the Arctic. Moreover, the Norwegian and US Defence Ministers pledged their support in providing Russia with technological and other assistance to help defuel nuclear submarines removed from service, and to develop safe storage facilities for spent nuclear fuel and radioactive waste. The AMEC Agreement establishes an institutional framework for contact and co-operation between military authorities in the three states. By 1998, however, the agreement had little to show for itself in terms of practical results.

In the spring of 1998, US authorities decided to link AMEC with the Nunn-Lugar Co-operative Threat Reduction (CTR) Programme (Sawhill 2000). The CTR Programme was created by the US Congress in 1991 as a mechanism to assist the Soviet Union in complying with its obligations of arms reduction in connection with the START-1 Agreement, and it is hoped, also new commitments under START-2.[7] It has provided more than US$2 billion to former Soviet states since 1991. One of the goals of the CTR Programme was to scrap thirty Russian ballistic missile submarines by 2001. The Russians currently have the capacity to scrap only a handful of submarines per year, the major obstacle being the defuelling process and dealing with the resulting waste and spent nuclear fuel. By linking AMEC and CTR, US authorities were able to provide a ready source of cash as well as indemnification from liability. Seven concrete projects have been identified under the AMEC co-operation so far, five of which are related to nuclear safety and are partly financed by the Plan of Action from the Norwegian side.[8]

The Nordic states in 1994 established the Nordic Environmental Finance Corporation (NEFCO), which has developed an environmental programme for Northwestern Russia. Several investment projects in the sphere of radioactive contamination from this programme are included in the Plan of

Action. At the initiative of Norway, a Contact Expert Group (CEG) on nuclear safety of radioactive waste management was established under the auspices of IAEA in 1996. The purpose of this initiative was to promote co-operation on projects to improve safety standards for the management, storage and disposal of spent nuclear fuel and radioactive waste in Russia. Also at the initiative of the Nordic states, the G-7 states established a Nuclear Safety Account within the European Bank for Reconstruction and Development (EBRD). The Nuclear Safety Account finances measures that contribute to technical and operational safety improvements in nuclear reactors in Central and Eastern Europe. Norway participates in and contributes financially to the Chernobyl Shelter Fund, which aims at building a sarcophagus around the Chernobyl nuclear power plant. Nuclear waste problems are also discussed in the North Atlantic Treaty Organisation (NATO), the North Atlantic Co-operation Council (NACC) and a Norwegian–French working group. Finally, Norway has taken a leading position in endeavours to create a Multilateral Nuclear Environmental Programme in the Russian Federation (MNEPR). The aim of this initiative is to secure satisfactory framework conditions for all participating parties in nuclear safety projects in Russia, e.g. related to indemnity against nuclear liability, access and oversight, and exemption from taxes, customs duties and other fees.[9] A Declaration of Principles was signed on 5 March 1999 (Ministry of Foreign Affairs 1999a), but a more binding legal framework is not in sight.

Industrial pollution in the European Arctic

Norilsk Nickel is one of Russia's leading producers of non-ferrous and platinum-group metals and the country's largest air polluter. Three of the company's six subsidiaries are located on the Kola Peninsula: the Pechenga-nickel Combine at Zapolyarnyy and Nikel,[10] the Severonickel Combine at Monchegorsk and the Olenegorsk Mechanical Plant at Olenegorsk. Pechenganickel and Severonickel emit large quantities of sulphur dioxide (SO_2) which causes considerable acid precipitation both on the Kola Peninsula and in the neighbouring Fenno-Scandinavian countries. This section reviews the problem of air pollution on the Kola Peninsula and the status of the area's mining and metallurgical complex.

The mining and metallurgical complex on the Kola Peninsula

RAO Norilsk Nikel was founded in 1994 by merging six companies under a corporate umbrella.[11] It has four main operation facilities: a mining and metal-processing facility at Norilsk in northeastern Siberia (see Figure 2.3); a metal-processing facility at Monchegorsk; mining and metal-processing facilities at Zapolyarnyy and Nikel; and a precious-metals processing plant at Krasnoyarsk in central Siberia. In addition, the company has a research institute (Gipronikel Planning and Design Institute) in St Petersburg and a

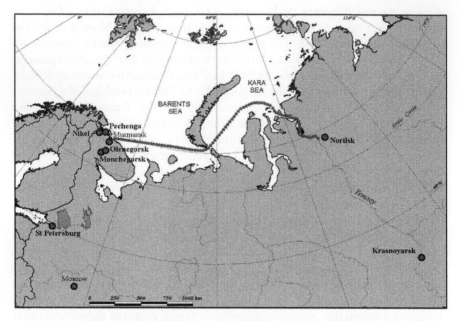

Figure 2.3 Location of the RAO Norilsk Nickel companies and shipment route
from Norilsk to the Kola Peninsula.

Source: Claes Lykke Ragner/FNI

mechanical plant at Olenegorsk on the Kola Peninsula. Norilsk Nikel is the
world's leading producer of nickel and palladium, and the company's sales
constitute approximately US$3 billion annually (Bond and Levine 2001).

The Pechenganickel Combine operates four small mines and a smelter in
the northern parts of the Kola Peninsula, close to the Norwegian border.
Mineral production in the region dates back to the early 1940s, when it was a
part of Finland. At the end of the Second World War, the Petsamo (Pechenga)
district was annexed by the Soviet Union, and mining and smelting operations were resumed at Nikel in 1946. In the mid-1960s, the focus of mining
activities was shifted to the Zhdanov deposit near Zapolyarnyy. Some 85 per
cent of the company's mine output comes from two open pits of this deposit,
scheduled to be worked out by 2005–2006. The remaining output comes
from three underground mines, which are projected for depletion at various
times during the period 2005–2015 (Bond and Levine 2001).

The metallurgical operations of Pechenganickel take place at the
company's plant at Nikel. The plant processes ore concentrates from the
mines near Zapolyarnyy and raw materials shipped over the northern sea
route (see Figure 2.3) from Norilsk. These shipments started in the late 1960s
when local ores began to decline. During the Soviet era, Norilsk Nickel shipped
approximately 1 million tonnes of ore from Siberia to its Kola facilities every

year (Kotov and Nikitina 1998a). The shift to a market economy has forced the company to reduce these shipments and rely more heavily on its deposits on the Kola Peninsula. Consequently, annual shipments have been more than halved since Soviet times (Kotov and Nikitina 1998a).

Pechenganickel sends its smelter output to the Severonickel Combine at Monchegorsk in the central parts of the Kola Peninsula. The combine has both smelting and refining facilities for processing nickel and copper. Production started in 1939, and since the late 1960s the combine has relied on non-local feedstocks, i.e. ore from Pechenganickel and Norilsk. Today, Severonickel operates Russia's largest capacity nickel refinery (Bond and Levine 2001).

Norilsk Nickel was privatised in 1993, although 51 per cent of the voting shares were to remain in government hands for another three years. However, Kotov and Nikitina (1998a) argue that the government soon lost actual control over the company as a result of the general chaos in the period, and that the company's directors, largely free from shareholder control, consolidated their hold on the enterprise and restructured it to further their own interests:

> In the chaotic period following the company's privatization, neither the Russian government nor the other nominal owners of Norilsk Nickel were able to exercise effective control over the enterprise. Largely unaccountable to anyone, Norilsk's management (the so-called red directors) ran the company for personal profit rather than long-term viability. Naturally, they were not interested in making capital-intensive investments to benefit the environment. This situation highlights the crucial distinction between Norilsk's leadership and its owners. In the West, ownership implies that a company's managers are ultimately accountable to the company's owners, and thus will pay attention to promoting that company's long-term interests. Managers whose primary goal is to increase the value of their company will make the investments (including environmental investments) necessary to enhance its prospects. In Norilsk's situation, however, exactly the opposite occurred.
>
> (Kotov and Nikitina 1998a: 565)

In recent years, Norilsk Nickel has increasingly oriented its production towards exports, which now account for more than 50 per cent of the company's sales and nearly 85 per cent of all Russian exports of nickel (Kotov and Nikitina 1998a). The company consolidated its economic position considerably during the 1990s. Like many other natural resource-producing companies in Russia, it got through the general economic crisis in the country fairly well. Since export taxes have become an important source of revenue for the Russian government, the authorities also have a vested economic interest in the company.

Pollution from the smelters

The Norilsk Nickel company towns have routinely ranked among the most polluted cities in Russia. Since the late 1980s, they have also become increasingly famous in the West as areas of environmental catastrophes. During the period from 1980 to 1987, the annual SO_2 emissions from Pechenganickel fell from 384,000 tonnes to 337,000 tonnes while emissions from Severonickel rose from 200,000 tonnes to 224,000 tonnes (Darst 2001). In 1994, Pechenganickel emitted 132,900 tonnes and Severonickel 111,000 tonnes of air pollutants (Kotov and Nikitina 1998a). In 2000, the corresponding figures were 160,860 tonnes and 57,397 tonnes (State Committee for Environmental Protection 2001).

The location of the Kola Peninsula smelters north of the Arctic Circle compounds the environmental problems as the Arctic ecosystems are more fragile and lack the assimilative capacity of those at lower latitudes. Hence, the activities of the Norilsk Nickel plants have led to wide-ranging environmental degradation and acidification. Close to the smelters, the forest is completely dead. According to the Arctic Monitoring and Assessment Programme's report on the state of the Arctic environment (AMAP 1997), the forest death area around Monchegorsk covers 400–500 square kilometres and extends 10 kilometres south and 15 kilometres north of the smelter complex. The zone is expanding at a rate of half a kilometre per year. The area severely affected by air pollution around Nikel and Zapolyarnyy is considerably larger than the one around Monchegorsk. It increased in size from 400 square kilometres in 1973 to 5,000 square kilometres in 1988 (AMAP 1997). The outer visible damage zone extends into the eastern parts of Inari in Finland. Also on the Norwegian side of the border, trees and other vegetation have been damaged (see Figure 2.4).

International regulation of industrial pollution

Related to the air pollution in the European Arctic, by far the most important international instrument is the UN Economic Commission for Europe Convention on Long-range Transboundary Air Pollution (LRTAP).[12] It addresses problems in Europe and North America concerning airborne pollutants, notably acid rain, and establishes a framework for co-ordinating pollution control measures and common emission standards.[13] The contracting parties are to take into account the precautionary approach as set forth in the 1992 Rio Declaration. They must reduce annual emissions from a reference year, and emission limits are established for some selected sources. A monitoring system has been set up, and five substantive protocols have been negotiated under the regime: on NO_X (1998), volatile organic compounds (1991), sulphur (1994), heavy metals (1998) and persistent organic pollutants (1998). The Soviet Union/Russian Federation has been an active partner in the LTRAP regime. Traditionally rather reserved towards

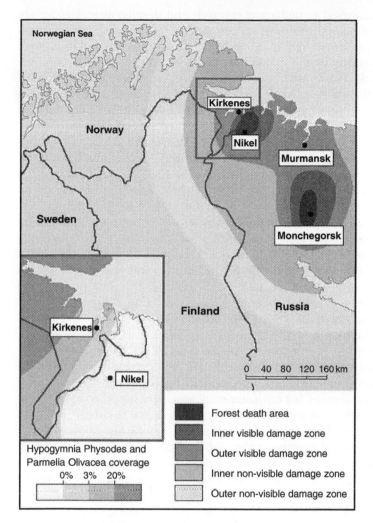

Figure 2.4 Approximate forest damage zones in the vicinity of Monchegorsk
and Nikel and the visible damage zones on the Kola Peninsula and
in Finnish Lapland.

Source: AMAP (1998).

co-operation with the West during the Cold War, in the late 1970s the Soviet
Union was enthusiastic in its support of the LRTAP process, regarding it
more in terms of 'high politics' than from an environmental point of view
(Kotov and Nikitina 1998a). At present, Russia has ratified the Convention
itself and the NO_x Protocol and signed, but not ratified, the Sulphur Protocol.

The Arctic Environmental Protection Strategy (AEPS), a programmatic,
legally non-binding document and process initiated by Finland in 1991,
commits the eight Arctic states (Russia, Canada, the United States and the

five Nordic countries) to undertake research and develop strategies for six priority environmental problems. A number of co-operative programmes have been established: Arctic Monitoring and Assessment Programme (AMAP), Protection of the Arctic Marine Environment (PAME), Conservation of Arctic Flora and Fauna (CAFF) and Emergency Preparedness, Prevention and Response (EPPR) programme. These programmes reported to the Ministers of the Environment of the Arctic countries, who in their turn identified priority areas for further action. Four ministerial conferences were held under the AEPS framework between 1991 and 1997. The AEPS programmes have now been subsumed under the Arctic Council, a forum established by the Arctic states in 1996. At the first meeting of the Arctic Council, a Regional Programme of Action for the Protection of the Arctic Marine Environment (RPA), developed by the PAME working group, was adopted. VanderZwaag (2000: 192) describes the adoption of the RPA as 'small, soft steps in addressing land-based pollution in the Arctic', observing at the same time that detailed actions are only recommended for POPs (persistent organic pollutants) and heavy metals, and that financial and technical commitments are left uncertain, although the need to assist the Russian Federation in taking pollution prevention actions is stressed.

Environmental co-operation between Russia and the Nordic states in the Barents Euro-Arctic Region (BEAR) and bilaterally is also of relevance for air pollution control in Northwestern Russia. The BEAR co-operation effort between Norway, Sweden, Finland and Russia was established in 1993 and has both national and regional components.[14] It includes co-operation in a range of functional areas, but environmental issues are supposed to permeate the regime as a whole. A separate working group on the environment is also part of the regime. Stokke (2000b) argues that the BEAR co-operation is closely linked to other regimes at the global, regional and bilateral level of relevance to the environment of the European Arctic. Of particular importance at the bilateral level is the Joint Russian–Norwegian Environmental Commission, which was established in 1988. Under the Joint Commission, comprising leading environmental agencies in the two states, several working groups have been established, including the Working Group on Airborne Pollution. The latter has developed an environmental monitoring and modelling programme for the border areas. The focus of capacity-enhancement efforts has been on the question of modernisation of the Pechenga smelter-work, which was raised to the governmental level when the Joint Environmental Commission was established. Norway offered to contribute 300 million NOK (at the time some US$50 million), but after years of planning the project was halted in 1997. A Finnish initiative was also stillborn. Contrary to expectations, the Norwegian project was revived in early 2001, and on 19 June that year the Norwegian Minister of the Environment and the Russian Minister of Economy signed an agreement on a modernisation project that would involve a 90 per cent reduction in emissions of SO_2 and heavy metals.

Conclusion

This chapter has given us an overview of some of the main environmental problems of the European Arctic and the international institutional arrangements aimed at their solution. Both the problems and the institutions vary in character. Industrial pollution and resource depletion are imminent problems – with the latter changing more over time than the former – whereas the nuclear complex of Northwestern Russia is problematic in terms of posing a threat of serious accidents rather than relating to current radiation levels in the region. The regime for fisheries management in the Barents Sea is a rather stable arrangement designed to take care of regulatory issues over a long period of time; some of the institutions set up to prevent industrial pollution and provide nuclear safety are more of an ad-hoc and programmatic nature. As the size of the fish stocks fluctuates, so does the reputation of the fisheries management regime. The achievements of the nuclear safety and industrial pollution control regimes are so far not evaluated too highly.

The most important fish stocks of the Barents Sea are definitely in a crisis at the turn of the millennium. Likewise, the environment of the European Arctic is being affected adversely by industrial activity in the region, and there is clearly a certain risk of nuclear accidents. However, the seriousness of these problems is a matter of interpretation. How we interpret – or 'frame' – the problems, also affects the choices we make regarding their solution. It is the process of 'framing' the relevant problems and linking problem perceptions to possible political solutions which is at the heart of the following discussion. A few practical questions stand out on the basis of the above presentation: why did the fisheries managers of the Joint Russian–Norwegian Fisheries Commission set the Barents Sea quotas far above the scientific recommendations towards the end of the 1990s? Why has there been such a massive international effort to contribute to nuclear safety in Northwestern Russia when the radioactive sources pose only a minor threat to non-Russian territories? Finally, why have Western governments been so keen on modernising the Kola smelters, and why has it been so difficult to get the Russians interested in these projects?

3 Discourses on marine living resources

The Norwegian party notes that the level of the cod quota is alarmingly high in consideration of available stock assessments and the recommendations from ICES. Taking into account the difficult conditions of the population of Northwestern Russia . . ., Norway has nevertheless found it possible to enter into this agreement.

(Protocol from the Twenty-eighth session of the Joint Russian–Norwegian Fisheries Commission)

Norway does everything it can to destroy the Russian fishing industry. And that's good. That's how it should be.

(Inhabitant of Murmansk)

Introduction

In Chapter 2, we noted that several of the most important fish stocks in the Barents Sea were judged by scientists at the turn of the millennium to be at risk of collapse. There is a corresponding tendency in the scientific literature to evaluate the achievements of the management regime negatively in the face of the declining fish stocks. While it is difficult to establish a causal link between choices made within the regime and the status of the regulated stocks, it would, from the point of view of discourse analysis, be of interest to elucidate the context in which important management decisions have been made. How did decision-makers explain their intentions behind the choices made? Why were some proposed solutions chosen and not others? How did the framing of the problem influence the selection among available 'solutions'? To what extent were choices determined by factors external to the regime? Whether the choices made actually contributed to the solution of the problem is an issue that lies outside the scope of this discussion.

The Russian–Norwegian regime for the management of the Barents Sea fish resources is responsible for the establishment of quotas on the basis of the scientific recommendations of the International Council for the Exploration of the Sea (ICES), for introducing regulatory measures related to fishing, and for overseeing Norwegian and Russian enforcement activities in the Barents Sea. A continuous discussion has taken place in recent years – in scientific and management circles, politics and society at large – as to what the 'right' quota sizes, regulatory measures and enforcement levels actually are. A striking feature of the management process towards the end of the 1990s was the increasing tendency of decision-makers to set total allowable catches (TACs) far above the scientific recommendations.

The main objective of this chapter is to investigate the discursive practices accompanying the scientific recommendations issued by ICES and establishment of TACs by the Joint Russian–Norwegian Fisheries Commission (hereafter referred to as the Joint Fisheries Commission) at the turn of the millennium. First, we ask how one defines what is to be counted as 'legitimate' knowledge in the formulation of scientific recommendations. How does ICES underpin its recommendations, and how is this advice perceived by the Joint Fisheries Commission and decision-makers in the national systems for fisheries management in Russia and Norway? Second, how is national interest intertwined in discussions about the establishment of TACs in the Joint Fisheries Commission? Third, how are the institutional arrangements accompanying the establishment of quotas perceived by the actors involved and by the general public in Russia and Norway? In short, how are the limits concerning what can be said and done in this particular discussion determined?

The chapter tells the story of how the Joint Fisheries Commission established its TACs at the turn of the millennium, i.e. mainly the years 1999–2001. While discursive practices surrounding the establishment of TACs represent the heart of the discussion, discourses related to other management practices, such as technical regulation of fishing activities and enforcement of rules, are referred to in the discussions of quota settlements, in order to supplement the argument. First, the relationship between scientific recommendations and the establishment of TACs in the Barents Sea cod fisheries during the latter half of the 1990s is investigated in some detail. Next, discursive practices related to knowledge, interests and institutions are presented with a view to outlining the empirical data rather than indulging in theoretical discussions. On this basis, the major discourses on marine living resources are defined and dissected from a more theoretical point of view. This section also discusses how discourses of a more overarching nature in society influence environmental discourse, in our case, discourse on marine living resources.[1] Finally, the genealogy of the discourse is summed up and some careful observations made regarding the relationship between discourse and power in the Barents Sea fisheries management.

Scientific recommendations and established quotas in the 1990s

Since the early 1960s, management recommendations on the Northeast Arctic cod stock have been given regularly by ICES through its Advisory Committee for Fisheries Management (ACFM). The first TAC for this stock was introduced in 1975; until then no effective management measures were in place for demersal fish (living close to the seabed) in the area (Nakken 1998). At the end of the 1970s, catches of cod and haddock were declining, and the rates of exploitation were far above advised levels. The Barents Sea capelin stock collapsed in the mid-1980s, which led to a lack of food for the cod and haddock stocks in the area. In 1990, the total yield from the area was at its lowest point since 1945 (Nakken 1998).

During the 1990s, the Northeast Arctic cod stock fluctuated from a state of near collapse at the beginning of the decade, through a period of intensive growth in the mid-1990s, and back again to near collapse at the end of the decade. As follows from Table 3.1, there was a tendency throughout the period to set higher TACs than advised by ICES. This was the case for the years 1992–1995 and 1998–2001; only twice – in 1990 and 1996 – were the TACs lower than the scientific recommendations. While TACs quickly increased to levels far above recommendations as soon as the stock started to recover (1992–1993), the quotas were set in the immediate vicinity of the recommended levels during the 'prosperous' years of the mid-1990s (1994–1998). When signs of crisis again emerged at the end of the decade, decision-makers did not take immediate action. The situation was particularly critical when the quota for 2000 was to be set in the autumn of 1999. ICES had provided an 'all time low' recommendation of 110,000 tonnes, down from a recommended 360,000 tonnes and a TAC of 480,000 tonnes for 1999. Only after a breakdown in negotiations lasting several days did the Joint Fisheries Commission agree on a TAC of 390,000 tonnes for 2000, i.e. a quota almost four times higher than the scientific recommendations. One year later, ICES increased its quota recommendations slightly, and the Joint Fisheries Commission for the first time agreed to set a three-year quota, at the level of 395,000 tonnes.

Table 3.1 Recommended and established total allowable catch (TAC) for Northeast Arctic cod during the period 1999–2001 (in 1,000 tonnes)

Year	Recommended TAC	Established TAC
1990	172	160
1991	215	215
1992	250	356
1993	256	500
1994	649	700
1995	681	700
1996	746	700
1997	750–850	850
1998	514	654
1999	360	480
2000	110	390
2001	263	395

Sources: Nakken (1998) and Aasjord (2001).

In addition to the discrepancy between recommended and established TACs comes the fact that the quotas recommended by ICES for the Northeast Arctic cod were for most of the years based on too high estimates of spawning stocks. The annual assessments for cod underestimated fishing mortality and overestimated stock numbers. In other words, fish were removed from the stock at a higher rate than the scientists bargained for on the basis of their analyses at the time the advice was given (Nakken 1998).

Discourse on knowledge

What role did knowledge play in the establishment of quotas in the Joint Fisheries Commission towards the end of the 1990s? From what we have seen so far, it looks as if scientific knowledge at least was of little importance; such advice was more or less systematically overlooked in the establishment of TACs for the Northeast Arctic cod stock during the years under investigation. Why was this? Were the figures perceived in some way as 'illegitimate' by Russian and Norwegian decision-makers in the Joint Fisheries Commission? Were other types of knowledge regarded as more germane than scientific knowledge? Or was the question of knowledge considered to be more or less irrelevant for the establishment of quotas?

The Joint Fisheries Commission's primary knowledge base is provided by ICES through its annual stock assessments and quota recommendations. This section reviews the principles underlying the provision of scientific knowledge, how these principles are perceived – or spoken about – in scientific circles and by non-scientific actors within and outside the Russian and Norwegian fisheries management systems.

The emergence of the precautionary principle

The leading international principle for fisheries management has until recently been based on securing the *maximum sustainable yield* (MSY) from fish stocks. MSY denotes the level at which the greatest quantity of fish can be caught annually without the total size of the stock being reduced. This principle is laid down in the 1982 Law of the Sea Convention (United Nations 1982), the major global arrangement related to the management of fisheries. Since the early 1990s, new concepts, such as *responsible management*, *sustainable development* and the *precautionary principle*, have come to the fore in international fisheries management. The precautionary principle was established internationally in the 1992 Rio Declaration:

> In order to protect the environment, the precautionary approach shall be widely applied by states according to their capabilities. Where there are threats of serious or irreversible damage, lack of full scientific certainty shall not be used as a reason for postponing cost-effective measures to prevent environmental degradation.
>
> (United Nations 1992: Art. 15)

The precautionary approach has subsequently been incorporated into international agreements pertaining specifically to fisheries management,[2] notably the UN Food and Agriculture Organisation's (FAO) Code of Conduct for Responsible Fisheries (FAO 1995) and the UN Agreement on Straddling and Highly Migratory Fish Stocks (United Nations 1995):

> States should apply the precautionary approach widely to conservation, management and exploitation of living aquatic recourses in order to protect them and preserve the aquatic environment. The absence of adequate scientific information should not be used as a reason for postponing or failing to take conservation and management measures.
>
> (FAO 1995: Para. 7.5.1)

> States shall apply the precautionary approach widely to conservation, management and exploitation of straddling fish stocks and highly migratory fish stocks in order to protect the living marine resources and preserve the marine environment.
>
> (United Nations 1995: Art. 6.1)

> States shall be more cautious when information is uncertain, unreliable or inadequate. The absence of adequate scientific information shall not be used as a reason for postponing or failing to take conservation and management measures.
>
> (United Nations 1995: Art. 6.2)

Hence, the essence of the precautionary approach is that lack of scientific knowledge should not be used as a reason for failing to undertake management measures that could prevent the degradation of the environment or the depletion of common-pool resources.[3] Whereas it was once considered reasonable to take such measures only when it was established with a high degree of certainty that the environment or resource basis would be seriously threatened without such interference, the introduction of the precautionary approach turned the proof burden upside-down: preventive measures should be postponed or omitted only when there was full scientific certainty that they were not necessary. The FAO Code of Conduct and the UN Agreement on Straddling and Highly Migratory Fish Stocks also pay particular attention to the implementation of the precautionary approach:

> In implementing the precautionary approach, states should take into account, inter alia, uncertainties relating to the size and productivity of the stocks, reference points, stock condition in relation to such reference points, levels and distribution of fishing mortality and the impact of fishing activities, including discards, on non-target and associated and dependent species as well as environmental and socio-economic conditions.
>
> (FAO 1995: Para. 7.5.2)

States and subregional or regional fisheries management organisations and arrangements should, on the basis of the best scientific advice available, inter alia, determine:

a) stock specific target reference points and, at the same time, the action to be taken if they are exceeded; and

b) stock specific limit reference points and, at the same time, the action to be taken if they are exceeded; when a limit reference point is approached, measures should be taken to ensure that it will not be exceeded.

(FAO 1995: Para. 7.5.3)

In implementing the precautionary approach, the following should, among other things, be considered:

6.4   States shall take measures to ensure that, when reference points are approached, they will not be exceeded. In the event that they are exceeded, States shall, without delay, take the action determined under paragraph 3 (b) to restore the stocks.

6.5   Where the status of the target stocks or non-target or associated or dependent species is of concern, States shall subject such stocks and species to enhanced monitoring in order to review their status and the efficacy of conservation and management measures. They shall revise those measures regularly in the light of new information.

(United Nations 1995)

The two 1995 global agreements relating to fisheries management, the FAO Code of Conduct for Responsible Fisheries and the UN Agreement on Straddling and Highly Migratory Fish Stocks, thus put emphasis on the interface between scientific uncertainties and the best scientific knowledge available. They prescribe the use of stock-specific reference points as a tool to deal with matters of risk and uncertainty in fisheries management. In Annex II to the UN Agreement on Straddling Stocks and Highly Migratory Fish Stocks, a precautionary reference point is defined as 'an estimated value derived through an agreed scientific procedure, which corresponds to the state of the resource and of the fishery, and which can be used as a guide for fisheries management' (United Nations 1995: Annex II, Para. 1). Further, the Annex states that two types of precautionary reference points should be used: limit reference points and target reference points. The former sets boundaries that are intended to constrain harvesting within safe biological limits within which the stocks can produce maximum sustainable yield. The latter reference points are intended to meet management objectives. Management strategies are supposed to seek to maintain or restore stocks at levels consistent with the agreed-upon target reference points and include measures that can be implemented when reference points are approached. It should be a goal for fisheries management systems to ensure that target reference points are not exceeded on average. The fishing mortality rate – or in non-scientific terms, the catch rate – that generates maximum sustainable yield should be regarded as a minimum standard for limit reference points. This implies that a management body that sets a limit fishing mortality above maximum sustainable yield would have to demonstrate why it is considered to be precautionary.[4]

ICES and the precautionary approach

In the autumn of 1996, it was decided at the Eighty-fourth Annual Science Conference of ICES to establish a Study Group on the Precautionary Approach to Fisheries Management (hereafter the Study Group). The mission of the Study Group was to draft a new form of ACFM recommendations incorporating the precautionary approach. In its first report, the Study Group stated that 'the adoption of the precautionary approach has consider-able implications for fishery management agencies and the fishing industry. It also provides an impressive list of tasks which the scientific community, in general, and ICES, in particular, needs to address' (ICES 1997: 4). Further, it states that the FAO Code of Conduct and the UN Agreement on Straddling Stocks and Highly Migratory Fish Stocks require the following technical developments:

1. the determination of reference points, with a priority for limit reference points that define the constraints on long-term sustain-ability, both in theory and as applicable to each stock;
2. improvements in the methods for dealing with uncertainties, notably in relation to evaluating the risk of either approaching or exceeding the limit reference points;
3. the evaluation of how well alternative harvest control rules either maintain stocks in, or restore them to, healthy states.

(ICES 1997: 5)

On this basis, the Study Group proposes that ICES should:

1. explicitly consider and incorporate uncertainty about the state of stocks into management scenarios; explain clearly and usefully the implications of uncertainty to fishery management agencies;
2. propose thresholds which ensure that limit reference points are not exceeded, taking into account existing knowledge and uncertainties;
3. encourage and assist fishery management agencies in formulating fisheries management and recovery plans. To do this effectively may require ICES to assist fishery management agencies in the develop-ment of coherent, measurable objectives;

    4. quantify and advice on the effects of fisheries on target and non target species, and on biodiversity and habitats;

    5. provide advice on fishing fleets and multispecies fisheries systems as well as on single stocks;

    6. evaluate fisheries management systems incorporating biological, social and economic factors as appropriate.

(ICES 1997: 5)

In its discussion about management implications of the introduction of the precautionary approach, the Study Group concludes that a change in culture is necessary within management agencies towards a management approach less focused on and influenced by short-term considerations and more with long-term sustainability. Further, the report of the Study Group elaborates the basis for calculations of concrete reference points for the various fish stocks within the responsibility area of ICES. The reference points are viewed as signposts for the provision of information on the status of the stocks in relation to predefined *limits* that should be avoided and *targets* that should be aimed at in order to achieve the management objectives.[5]

In 1998, ICES proposed the reference points for the Northeast Arctic cod stock as shown in Table 3.2.

B refers to the size of the spawning stock while F indicates fishing mortality or catch rate. The $-_{lim}$ values represent levels of spawning stock and catches associated with potential stock collapse ('limits to be avoided'); the $-_{pa}$ values imply precautionary approach levels for the spawning stock and catch rate ('targets to be aimed at'). A catch rate higher than F_{pa} is defined as over-fishing; a stock with a spawning stock biomass lower than B_{pa} is considered to be over-fished. Hence, the management of the Northeast Arctic cod stock is held to be in accordance with the precautionary approach only as long as the stock's spawning mass is larger than 500,000 tonnes and the catch rate is lower than 0.42. The reference point for the spawning stock size of 500,000 tonnes equals the 'safe biological limit' applied by ICES for this stock since 1986 (Aasjord 2001).

Table 3.2 ICES reference points for the Northeast Arctic cod stock

ICES considers that	ICES proposes that
B_{lim} is 112,000 t, the lowest observed in the 53 year time series	B_{pa} is set at 500,000 t, the SSB* below which the probability of poor year classes increases
F_{lim} is 0.70, the fishing mortality associated with potential stock collapse	F_{pa} be set at 0.42: this value is considered to have a 95% probability of avoiding the F_{lim}

Source: ICES (1998b).

Note: *spawning stock biomass.

In its last assessment report on the Northeast Arctic cod stock, the ACFM concludes:

> The stock is outside safe biological limits. Fishing mortality in the last four years has been among the highest observed and well above F_{pa}, even above F_{lim}, and is not sustainable. SSB has been below B_{pa} since 1998. Surveys indicate below average 1998 and 2000 year classes and a very poor 1999 year class.
>
> (ICES 2001: 2)

Norway, Russia and the precautionary approach

In Norway, the principles of sustainable development, responsible management and the precautionary approach are incorporated into the country's official fishery policy:

> Our ocean areas represent large resource and environmental values. Norway has a responsibility to ensure a *sustainable* management of these areas. . . . An overarching goal for the resource management is to regulate harvests so as to secure a *responsible* exploitation of the resources. . . . Inadequate knowledge on the resource basis and the marine environment demands margins of safety in their management. *Inter alia*, in accordance with international agreements the *precautionary* approach shall be the basis for Norwegian resource management.
>
> (Stortinget 1997: 15, emphasis added)[6]

Russia has no federal law on fisheries as yet,[7] and it has not been possible to find reference to any of these concepts in the jungle of normative documents issued at lower levels of the legal hierarchy in the area of fisheries, nor in the Law on the Exclusive Economic Zone of the Russian Federation, adopted in 1998 (Russian Federation 1998). The latter instead speaks of principles such as 'rational use of marine bio-resources' (Russian Federation 1998: Chapter 2) and 'protection of the marine environment' (Chapter 5) as separate entities. On the other hand, Russia has ratified the UN Agreement on Straddling and Highly Migratory Fish Stocks (United Nations 1995), where the precautionary approach was first established as a fundamental principle for fisheries management.

In the protocol from its 1997 session, the Joint Fisheries Commission noted:

> The parties agreed on the need to develop further long-term strategies for the management of the joint stocks of the Barents Sea. Until such a strategy is available for cod, the parties agreed that the annual total quota is to be established so as the spawning stock is maintained above

500,000 tonnes at the same time as the fishing mortality in the next years is reduced to less than F_{med} [safe biological limit] = 0.46.

(Ministry of Fisheries 1997: 2)

The same paragraph is used in the protocol from the 1998 session, with the specification that fishing mortality shall be reduced to less than 0.46 'no later than in 2001' (Ministry of Fisheries 1998: 2). In the protocol from the 1999 session, F_{med} (safe biological limit) is changed to F_{pa} (precautionary approach) and the aimed-at catch rate level reduced from 0.46 to 0.42 (Ministry of Fisheries 1999: 2), i.e. brought into accordance with the recommendations from ICES. In 2000, the Joint Fisheries Commission requested ICES to 'reconsider the B_{pa} in light of the dynamics of the cod stock over the last 30–40 years' (Ministry of Fisheries 2000: 2). This replicates a unilateral Russian request in a letter to ICES of 7 August 2000 (ICES 2000). Although the wording of the letter urges ICES to 'reconsider' the reference point, it is clear that both the Joint Fisheries Commission and the Russian Government are in fact calling for a reduction. ICES has so far declined to reduce the reference point, referring to ongoing analysis that may in the future lead to such a change (ICES 2001).

How the scientific knowledge is perceived

Norway has explicitly embraced the precautionary approach as the basis for the country's fisheries management. The same can be said about the Joint Fisheries Commission as a result of its adoption of the reference point B_{pa}. The principle is not reflected in Russian legislation at the level of law, but Russia has accepted it through its ratification of global agreements and participation in ICES and the bilateral regime with Norway. Despite this formal embracement of the precautionary principle, it is difficult to see that the Joint Fisheries Commission has complied to any significant extent with ICES demands that Barents Sea cod fisheries shall be harmonised with the principle. As said, the TAC for cod was set higher than recommended by ICES in nine out of ten years in the period 1992–2001. But more specifically in relation to ICES's precautionary reference points, fishing mortality has been higher than 0.42 – i.e. the stock has by definition been over-fished – in seventeen of the twenty years 1982–2001. In fifteen of these years, fishing mortality was higher than 0.7, i.e. the catch was at a level considered to represent a risk of stock collapse (Aasjord 2001). How do the participants in the regime explain the discrepancy between established principles and political practice? And how do the targets of fisheries management regard the research activities and scientific recommendations that underlie the establishment of quotas?

In interviews with Norwegian and Russian fishers (i.e. captains on fishing vessels) during the period 1997–1998 carried out for a study on compliance in the Barents Sea fisheries, I concluded that the Norwegian fishers seemed to have a deep-rooted scepticism regarding the work of marine scientists while the Russian fishers revealed a more positive attitude (Hønneland 2000a). The following interview extracts were typical of the investigation:

> The researchers don't even believe in the results themselves. It's impossible to plan what we can fish next year. . . . It deprives us of the possibility of planning. When we receive wrong signals, it's not easy. . . . The poor planning can be the one thing that makes *me* become one of the violators.[8]

> My confidence in marine science is bloody low. It's impossible to run a business in a reasonable way. It's an extremely important thing they're left in charge of, . . . and they keep messing around in the Barents Sea with a couple of vessels, it's ridiculous! I don't think they have *ever* been right [in their estimates]! They should rather put people on board the Coast Guard vessels . . ., or alternate between fishing vessels. This year, it's been an extremely big problem; everybody has a problem with it. I don't think there's much respect for the marine scientists among fishers; people sit in Tromsø and give orders; ...and the refusals from the Russian zone[9]. . ., it just makes me laugh. There won't come anything sensible out of it anyway. That's the kind of things that make people violate the law. This year, saithe quotas are historically low, and there's so much saithe in the sea you can't hide from it! The saithe is a bloody fast swimmer, and we can't keep the bloody trawl net away from it! It's the same with the Greenland halibut: we bloody burst the cod-line with it! And there's supposed to be little of that sort! I tell you: there's excessively much saithe in the sea! We made a budget for 50 tonnes and have already taken 170 so far this year; and there's supposed to be no saithe! They should have used the Coast Guard; they're on the fishing grounds where things happen. There's an extreme arrogance among the marine scientists. They don't listen to the Coast Guard either. . . . You lose respect for the quotas that are established. . . . It's so bloody insane that I've considered giving up fishing.[10]

> Of course, it's necessary to do research on the fish stocks. . . . The specialised research vessels do a very good job, but commercial vessels with researchers on board don't. Their main task is to fish. The specialised research vessels do what they're supposed to do; they're important, necessary.[11]

> I have a positive attitude towards [the marine scientists]. The Barents Sea is not inexhaustible. It's quite right that the quantity of fish is checked. The fish is to be preserved for future generations. I'm not competent to assess whether the estimates are right; this is a specific science.[12]

The difference between Norwegian and Russian fishers was striking. One possible explanation for their different attitudes to marine scientists might be the traditionally strong position of science, and particularly the natural sciences, in Russia. While there seems to be a considerable distaste of scientific experts among Norwegian fishers, this does not seem to be the case with Russian fishers. Another, though related, explanation is the fact that most Norwegian fishers are typical representatives of a periphery culture with long traditions of opposing 'everything which comes from the centre'. The homeport of the Russian fishers interviewed, Murmansk, is, on the other hand, an 'artificial' Arctic city in the sense that it has been built up during the course of the twentieth century. Most Russian captains in the Barents Sea fisheries originally hail from further south in the former Soviet Union and do not share the same sense of belonging to a periphery as the Norwegian Barents Sea fishers.

However, we do not find this difference between Russian and Norwegian actors in the discussions following the establishment of quotas for the Barents Sea cod stock for the years 1999–2002. [13] In fact, the scepticism to scientific opinion might, at least at first glance, seem to be stronger on the Russian side than in Norway.[14] A few extracts from media reports illustrate this:

'The scientists are totally on the wrong track as far as the size of the cod stock is concerned. We could easily have set a quota of 520,000 tonnes,' says [Russian shipowner], who is far from content despite the fact that the cod quota was once more set far above the scientific recommendations. . . . 'Contrary to the scientists, who go out at sea a couple of times a year, we follow activities at sea every day throughout the year. We know what we are doing!' . . . 'Catch reports for this year show good concentrations of cod in wide ocean areas. The size of the cod is not bad at all – at least compared to 1998. When I have recommended a TAC of 520,000 tonnes for next year, it is based on these observations. We trust them more than we trust the figures that ICES delivers.'[15]

Everyone who works at sea can see what's going on and that the scientists are wrong. I think Norwegian and Russian fishers should talk together ahead of the [Joint Fisheries] Commission meeting in Tromsø. The 2001 TAC for cod should be at least 600,000 tonnes. . . . We who have several dozens of trawlers in operation every day throughout the year know best how conditions are in the sea.[16]

'The cod stock in the Barents Sea is at the same level as in 1997. The only difference is the quotas, which are only half of the 1997 level,' says general director Nikolay Tropin in the Union of Private Fishery Enterprises in the North in Murmansk to the Internet newspaper *Rybnyy Murman*. . . . Tropin senses 'bumps in the ocean' of fish of all sizes: cod, halibut, herring and other species. None of the figures that the Norwegian Institute of Marine Research and the Russian Polar Institute come up with fit in for Tropin, who disagrees sincerely with both

Norwegian and Russian scientists. The union [he represents] organises hundreds of small shipowners in Murmansk and its vicinity. 'Reports we receive from our vessels indicate that all fish stocks in the Northeast Atlantic are [at the moment] under-exploited.'[17]

Critical views on the scientific recommendations are also found among Norwegian fishers and representatives of their organisations:

While scientists exaggerate the extent of the crisis, the fishing fleet is getting the best daily catches in years. . . . What surprises the fishers most, is that there have never been such large amounts of accessible fish of exactly those species that the scientists can hardly find at all. It seems as if the stock assessments are built on a totally shaky foundation. [Addressing the regulators:] Use your common sense, not the scientific recommendations![18]

Almost to the day ten years ago, the same thing happened: the scientists found no cod and hence [it was established as a fact that] there was no cod in the Barents Sea. . . . Everyone with some knowledge about our fisheries knows what happened the following two to four years. Never had the cod been so easy to catch, and never were there such large amounts of them. In my opinion, the arrogance displayed by the scientists at the time was unique.[19]

It's a fact that our trust in the Norwegian marine scientists, and in particular the ACFM, has evaporated. . . . The establishment of quotas for Northeast Arctic cod reveals that neither Norwegian nor Russian authorities have much confidence in the scientific recommendations. At the same time, it's a paradox that we neglect the scientific recommendations for stocks that they probably have quite a good overview of while we follow the recommendations nearly slavishly when it comes to stocks that they have less knowledge about, for instance saithe, red fish and halibut. . . . In my opinion, it's meaningless to establish constant reference points for catch rates and spawning stocks for a predatory fish like cod and then call it precautionary management. It is, in fact, quite the opposite. We cannot regulate the cod as a plankton eater, as we do today, but must take more into account the presence of [living] food [for the cod] in the sea when we set quotas.[20]

How does one get round these [scientific] models? The scientists will soon be as 'clever' as the fishers themselves with figures. How do they reach these figures? . . . There's no lack of excitement in our lives. . . . In 1990 . . ., the scientists said there was a resource crisis in the sea, but the fishers observed as much fish as they had ever seen before. [Now], they say that the [saithe] stock is on the way down and they reduce the quotas. At the same time, we constantly observe a considerable amount of saithe. . . . Where do the scientists get their analyses? . . . This is a challenge for the scientists: explain this to us![21]

On the other hand, there is a growing chorus of voices critical of the practice of neglecting scientific opinion, at least on the Norwegian side. They can be heard in the fishers' own organisations, in environmental non-governmental organisations (NGOs) and among the public at large, e.g. in the press. As expressed in an editorial published in the largest newspaper in Northern Norway on the eve of the Joint Fisheries Commission's session in 2000:

> Today, the results of the Norwegian–Russian [negotiations] will emerge from a negotiation climate, which, seen from outside, is far milder and less troubled than the sea from which the fish is to be caught. Nevertheless, there are stormy clouds over the Barents Sea. These clouds come not from Our Lord, but from those who have been given the responsibility to lay the foundations of a management system which should also take into consideration those generations of fish that have not yet been hatched and those generations of people that have not yet been born. Such a long-term perspective seems only partly to have guided [the decisions made by the Joint Fisheries Commission]. . . . This is gambling with the resource base, and Norway should oppose it with everything we have of expertise and weight. . . . Instead of being a good and safe guarantor of continued life both in the sea and ashore, politics has become perhaps the most dangerous enemy of life. Politicians, whether Russian or Norwegian, let themselves be influenced by a trawler fleet that is far too big and far too effective and over the years has showed itself incapable of taking responsibility.[22]

A couple of editorials from the Norwegian fisheries press illustrate the same point:

> We have three tough years ahead of us in the cod fishery in order to get stocks back up again. The reason for [this miserable situation] is largely due to the fishing of young specimens, 3- and 4-year olds, the age groups that are supposed to bring the spawning stock up to the minimum level of 500,000 tonnes in the years to come. Last year's recommendation from ACFM was 360,000 tonnes while the quota was set at 480,000 tonnes. This year's recommendation from the International Council for the Exploration of the Sea in Copenhagen, implying a drastic reduction in quota levels, therefore came as no surprise since the objective is still to maintain a spawning stock of 500,000 tonnes. Voices urging precaution in the fishing of cod in the Barents Sea have been unanimous since the meeting of the [Joint] Norwegian–Russian Fisheries Commission in autumn 1997. At the time, the parties agreed on a gradual reduction in quotas to ensure a sustainable spawning stock in 2001. This strategy has

not succeeded, partly because the quota was set too high by the Commission for the years 1998 and 1999, and, not least, because fishing over the last one and a half years has concentrated on young and first-time spawning fish, that is: the future spawning stock. Of course, this cannot continue.[23]

Next year's cod quota does not create much enthusiasm among fishers. Many feel, quite rightly, that a quota of 390,000 tonnes Northeast Arctic cod balances on a knife edge of what can be characterised as sustainable.[24]

Criticism of management is also emerging from the fishers themselves and their organisations on the Norwegian side. Notably, while they reiterate their scepticism concerning the scientific recommendations, they nevertheless say that quotas should be based on such recommendations if the processes underlying the recommendations could be improved:

'The fishers' criticism of the scientists has been very severe. Maybe we demand the impossible of the scientists today. The most important job of our organisation in the time to come will be to create better working conditions for the scientists,' says [deputy leader of the Norwegian Association of Fishers], who is not overly enthusiastic that next year's cod quota is set at 390,000 tonnes.[25]

We must not end up with a situation where we blame the scientists [for the miserable state of the cod stock]. If we really want to be world champions in fisheries management, then we must give more money for research. The research that takes place today simply cannot be good enough and we have to acknowledge that.[26]

The scientists themselves share the view that their recommendations have not always been optimal, but say that the tendency has been more to over-estimate the cod stock than the opposite:

The most important lesson after twenty years of catch quotas should be that the marine scientists almost systematically have overestimated the cod stock, and that the regulators must take this fact into consideration until we find out why this happens.[27]

Disagreement between Russian and Norwegian scientists participating in ICES and the bilateral regime seems to be marginal and related more to the interpretation of the data than to their substance.

We more or less agree on the figures of the cod stock, but there are diverging views [among Norwegian and Russian scientists] about how the figures should be interpreted and which regulatory measures they should lead to. The Russians want to interpret the insecurity in only one [optimistic] direction while, from the Norwegian side, we found it necessary to bring in factors that unfortunately go in the opposite direction. . . . The Russians' hopes for the cod stock are based on, among other things, positive outlooks for the marine environment and the growing amount of capelin in the sea. From our side, we had to emphasise that we have 'always' overestimated the stock and underestimated fishing mortality, that the last age groups are weak, and that the oceanographers predict lower temperatures in the sea in the years 2002–2004.[28]

We completely share the views held by Norwegian scientists. There is no major disagreement between us. We work together in ICES, we conduct joint scientific expeditions, and we are all behind the stock assessments and quota recommendations that ICES puts forward. What we do not accept, is the spawning stock reference point of 500,000 tonnes. This should be set far lower, at approximately 250,000–300,000 tonnes.[29]

According to a Norwegian scientist who has co-operated with Russian scientists for decades, a difference in opinion is discernible among Russian scientists, notably between the 'operative' researchers at the PINRO institute in Murmansk and their 'superiors' at the federal marine research institute in Moscow.[30] While the former by and large share the views of their Norwegian colleagues, leading Moscow-based scientists have in recent years taken the quite radical (or, in effect, rather old-fashioned) view that it is mainly the marine environment (e.g. water temperature) that influences on the size of the cod stock, not fishing mortality or catch levels.[31] This last position implies that the size of the quota will be of little importance; human activity will in any event not be the decisive factor in determining the size of the future cod stock. Now, although this position has apparently not been set out by Russian scientists before the Joint Fisheries Commission, the 'scientific views' of the Russian delegation on the Commission are based on the opinions of mainly Moscow-based scientists who, on other occasions, have endorsed this view.[32]

In sum, there is a considerable scepticism among Russian and Norwegian fishers, and, to some extent, also among regulators of the two countries, towards the methods and the results of marine science within ICES. A recurring aspect is the assumption that the fishers (and, implicitly, at least some shipowners) themselves are closer to the problem than the scientists and therefore in a better position to have a qualified opinion about the state of the fish stocks. Scientific opinion is subsequently considered to be fundamentally flawed and calls for the inclusion of 'traditional knowledge', i.e. practical knowledge gained by fishers, in the regulatory process. On the other hand, there is a growing public concern in Norway because the scientific advice is not being adhered to in the establishment of quotas. Environmental NGOs, media commentators, politicians and others criticise the decisions of the current management system. There is obviously a more positive view of scientific knowledge among this section of society. Some Norwegian fishers have joined in the chorus of critical voices, as mentioned previously. Workers in the Russian fishing industry and fisheries management system seem to have a stronger basic trust in the potential of marine science to produce guidelines for sound fisheries management. However, there is disagreement about the concrete reference points and the stock assessments and quota recommendations produced by ICES for the Northeast Arctic cod stock. Within the Joint Fisheries Commission, the Norwegian delegation seems to have partly supported the scientific opinion expressed by ICES while the Russians have opposed it. Hence, distrust in the scientific recommendations – or the premises underlying the existing system for the production of such recommendations – may be one reason why the Joint Fisheries Commission tends to set quotas far above the scientific recommendations. In particular, the distrust of the Russian delegation to the Joint Fisheries Commission may well have kept quota levels above ICES's recommendations. On the other hand, the distrust of scientific knowledge is not unanimous – not even among fishers – and there is growing public concern, at least in Norway, over recent developments. Hence, it would be rash to conclude that the political outcome in this case can be explained by a lack of legitimacy of scientific data among decision-makers.

Discourse on interests

This section focuses on how concern about interests has loomed in the debate on the establishment of the Barents Sea cod quotas. After an outline of interest discourse in Norwegian and Russian fisheries, we discuss in greater detail a concrete episode – the Norwegian Coast Guard's arrest of the Russian trawler *Chernigov* in the Fishery Protection Zone around Svalbard in April 2001 – and the public debate that ensued. Strictly speaking, this incident could have been included under the discussion of Russian (and Norwegian) interest discourse, but it touches upon other aspects of the Barents Sea fisheries management than the quota establishment and hence is treated separately.

Discourse on interest will inevitably involve the discourse on knowledge as far as the management of the Barents Sea fisheries is concerned. This section shows, among other things, how actors use knowledge-related arguments to promote the interests of specific groups.

The Norwegian discourse: battles between interest groups at the national level

The Norwegian fisheries discourse is basically a continual discussion about how the Norwegian share of the Barents Sea quota should be distributed among various groups within the Norwegian fishing fleet. The Norwegian fishery sector consists of a large number of actors characterised by highly divergent interests.[33] The main groups include the ocean-going fishing fleet, the coastal fishing fleet and the land-based fish-processing industry. The ocean-going fleet is made up of a relatively limited number of vessels. The most advanced of them, such as the factory trawlers, are mainly registered in the southwestern part of Norway. The coastal fishing fleet consists of a large number of small vessels fishing with conventional gear. Most of these boats are registered in Northern Norway. This fishery was not subject to quota limitations until 1989–1990. The fish-processing plants ashore constitute the economic backbones of many small fishing communities along the coast of Northern Norway. According to a rough estimation, some 10,000 fishers and 5,000 employees are involved in the fish-processing industry in Northern Norway (Hoel 1994). The three northernmost counties – the area usually referred to as Northern Norway – have a total population of approximately 460,000 people.

Quotas are shared among fishing vessels by the Regulation Council, on which sit fishery authorities, representatives of marine science, fishers' associations, trade unions for workers in the fish-processing industry and other organisations. The main conflicts pattern is between the coastal and the ocean-going fleets. Around 70 per cent of the cod quota has been allotted to the coastal fleet in recent years. The Regulation Council's job is to propose a distribution system for the Norwegian share of the joint quota. Final decisions are made by the Ministry of Fisheries. The policies implemented could be understood as 'a compromise between what can be defended biologically, legitimized politically and accepted on social and economic grounds' (Hoel *et al.* 1996: 293).

The Norwegian Fishers' Association is an important actor in the fisheries management system in Norway. It represents the vast majority of fishers on both ocean-going and coastal vessels and includes the Norwegian Association of Shipowners, which mainly represents the ocean-going fleet. Not only is it represented on the Regulation Council and the Joint Fisheries Commission, but also it maintains a continuous dialogue with the fishery authorities on current issues in the fisheries management process. A small group of coastal fishers have elected to stand outside the Fishers' Association. They are organised in the Norwegian Association of Coastal Fishers. In 1994, the Norwegian Fishers' Association adopted what is known as the 'cod ladder', which roughly means that ocean-going vessels will be allotted around 35 per cent of the total Norwegian cod quota in years with high quotas and approximately 25 per cent in years with lower quotas. It is an underlying premise in Norwegian fisheries management that the coastal fleet has a 'first right' to the fish; in years with an abundance of cod in the Barents Sea, the ocean-going fleet will see its share increased as compared to less prosperous years.

As we have seen, differing views were expressed by Norwegian fishers and their organisations about the way management was practiced in recent times. The Norwegian Fishers' Association has little trust in the scientific recommendations, but gives its support to the official view of the Norwegian delegation to the Joint Fisheries Commission, which expresses concern about the disregard shown to scientific recommendations in recent years. The Norwegian Association of Shipowners, for its part, is more outspoken in its criticism of the scientific recommendations and less concerned about the failure to reduce the quotas in accordance with recommendations from ICES. Finally, the Norwegian Association of Coastal Fishers takes the opposite position, expressing deep concern about the declining cod stocks and the consequences this could have for coastal communities. This last position is largely supported by environmental NGOs and parties on the political left. The following quotes give a further illustration of how discourses on knowledge and interest are intertwined in the Norwegian discourse on management of the Barents Sea fisheries:

I would like to praise the scientists for being firmer in their assessments than they have been earlier. This does not mean that I think that everything they do is right. It's a disaster for Norway generally and for coastal communities in particular that the cod quotas are being reduced, and when disasters happen a board of inquiry is usually set up to find out what went wrong. I think it's time we set up a public board of inquiry to find out what went wrong [in the management of the cod stock] because when we do not know what went wrong, we could easily find ourselves in the same situation once more. Is that what we want? Such a board of inquiry should in my opinion look at whether the politicians have been irresponsible in setting the quotas higher than the scientific recommendations. . . . It should investigate whether the concentration on the ocean-going fleet has been too heavy and whether capitalist considerations have preceded considerations of the marine resources. . . . Finally, I would like to say that I hope the politicians do not set quotas higher than the scientific recommendations. . . . I would like to continue to live in the periphery, but it must be in a strong and lively community, not in a dead one.[34]

'We must cut half of the trawlers from the fishery. Let's make nails out of them! For a period now, we must bear over with the cries of distress from shipowners and bank directors echoing in the mountains,' says Eirik Falch, leader of the Norwegian Association of Coastal Fishers. 'It's impossible to defend the capital that has been invested in the trawler fleet [and combine it] with a decent and sustainable fishery. The result is ruthless exploitation. The only way to do anything about this situation is to reorganise the fleet. The ocean-going fleet must be cut, and the fishing methods must aim for sustainability.'[35]

'Gross irresponsibility!' This is how the [co-ordinating committee for small-scale coastal fishers] characterises next year's cod quota. 'The cod quota for 2000 is set far above what all available data say is sustainable.'[36]

The results of the quota negotiations are clear: the quotas had to be set at four times above the scientific recommendations. Apparently they have to be this high to sustain living communities along the coast. The fishing industry – the processing factories and the coastal fishers – needs something to live from, not to speak of the ocean-going fleet. But what about next year? If the fishers take out quotas four times higher than the scientists recommend in order to maintain a sustainable development, a natural consequence could be that next year's recommendations will be even lower. Should we disregard these also? We cannot disregard the scientific opinion anymore. In order to ensure future catches of fish, we have to fish less now! It won't help saving the fishing industry and the small coastal communities this year if they are left without fish next year. As I see it, there is only one way out: we have to follow the recommendations given by the marine scientists! The quotas will be low, and we must allocate them to the coastal fishers. . . . The ocean-going fleet will simply have to grant itself one year's leave.[37]

Hence, the arguments produced in the Norwegian discourse on interest about the management of the Barents Sea fisheries can be divided into two main groups. First, there is the 'official' debate about the short- versus long-term interests of the Norwegian fishing industry and coastal communities. When representatives of the Norwegian fisheries management system or the Norwegian Fishers' Association comment on the outcomes of sessions in the Joint Fisheries Commission, they express concern about the 'long-term sustainability of the cod stock'. In other words, they view the alternative choices of the Joint Fisheries Commission as between short- and long-term yields from the fish stocks. The official Norwegian view is to give priority to long-term interests whereas the Russians are considered to think more in a short-term perspective. Second, an interest discourse is discernible related to the distribution of resources at the national level in Norway, reflecting in turn political discourses of a more general nature in society. Lately, this discourse has above all centred around the concrete issue of whether the Barents Sea cod quotas should follow the advice of marine scientists or not, revealing different positions with regard to the wider questions related to quota distribution or politics at large. We have seen that coastal fishers support a precautionary approach in the establishment of quotas, arguing that scientific recommendations should be followed and that the self-indulgence witnessed by the 'ruthless exploitation' in recent years should come to an end. At the same time, they also speak of 'too heavy a concentration on the ocean-going fleet', the failure resulting from 'capitalist considerations' and the 'cries of distress from shipowners and bank directors echoing in the mountains'. In other words, they are promoting the interests of the group of fishers to which they belong themselves. Moreover, they link issues of knowledge and conflicts of interest between groups of fishers to the wider political goal of sustaining a spread community structure in the country's northern periphery. A 'discourse coalition' (see Chapter 1) is emerging uniting Norwegian marine scientists, coastal fishers, environmental NGOs and the political left. We will return to this in the section on what I have chosen to call the 'sustainability' discourse.

The Russian discourse: the battle between Russia and the West

The Russian discourse on quota settlements in the Barents Sea is conspicuously void of allusions to interest conflicts among various groups within the country.[38] Likewise, discussions about the need to choose between short- and long-term interests are also practically absent. Instead, there is a clear tendency in Russian fishery circles to speak of the battle of interest between the two states involved, Norway and Russia.

First, the Russian delegation has apparently been quite unambiguous in emphasising at sessions of the Joint Fisheries Commission that it is not ready to reduce the year-on-year cod quota any further.[39] (This spurred a Norwegian journalist to refer to the Russians as *no-fish-Ivan.*)[40]

> There was a meeting in the regional administration with participants from the fishing industry and scientists from PINRO where . . . the tactics and strategy of the Russian party [to the Joint Fisheries Commission] were discussed. The principle that 'ours' were to follow in the establishment of TACs for cod and haddock was adopted unanimously: not give in [to the Norwegians] on a single kilo.[41]

Second, the Russians seem to perceive Norway as a rational, unitary actor with mainly economic motives in the establishment of TACs for the joint stocks in the Barents Sea. The Norwegians' insistence on internationally accepted principles for fisheries management, such as sustainability and the precautionary principle, are perceived to be a smoke screen hiding their real intentions.

> Norway has a very rational administrative system situated in a stable political environment. Before last year's quota negotiations, the Norwegians calculated exactly how large a loss their fishing industry could bear, made plans for compensating losers with revenue from the aqua-culture sector, and decided to go in for a reduction in the cod quota at this level.[42]

> Right from the start, the Norwegian delegation pursued a hard line based on ICES's recommendations which, to put it politely, are 'a bit

more precautionary than necessary' as far as the assessment of the cod stock is concerned. Here, it should be observed that a range of experts do not exclude the possibility that one of the factors behind the stipulation of these recommendations, with all due respect for this indisputably very respectful organisation, have been the interests of Norway and the EU countries, whose representatives constitute the majority of the members of ICES's working groups. And it is quite possible that the Norwegian delegation's strong demands for reductions in the cod catches are aimed at maintaining the high price of the country's fish export commodities. . . . This is the third year in a row that the Norwegian party attempts to achieve a reduction in the cod quota, and it is absolutely possible that this insistence is based on the fact that Norway has started artificial breeding of cod. In two to three years' time, the quantity of this fish may reach 180,000–200,000 tonnes a year, that is, practically equivalent to their share of the quota of 'wild' cod. And in order to maintain the price for the future, they wish to 'freeze' this amount of catch. Whether one likes it or not, the Norwegians at the session [of the Joint Fisheries Commission] gained additional support for their argument by the demands of the 'greens' and parts of the local press.[43]

There is also another reason why the Norwegians went in for the establishment of a fixed cod quota for three years ahead. The Russian participants to the session were told this behind the scenes [of the negotiations]: it is expected that in the course of exactly the next three years Norway will succeed in breeding amounts of cod that, if achieved, will make it possible to reduce the quota of this stock sharply.[44]

This way of thinking on the Russian side has been picked up by the Norwegian party:

We are witnessing an agitation in Russia which aims at building up a substantial opinion against Norway and ICES. We are accused of having a hidden agenda.[45]

We only want the best, but Russian suspicion [of the West] is still alive. They think there's always something behind our words and deeds – some kind of hidden agenda.[46]

Third, there is direct reference by the Russians to (what they perceive as) the decisive economic calculation of Norway as a 'natural' thing. The first quote below is the continuation of the extract above that depicted Norway as using ICES for its own economic motives.

[I]t is quite possible that the Norwegian delegation's strong demands for reductions in the cod catches have as their objective to maintain the high price of the country's fish export commodities. In principle, there is nothing special about this – every country defends its own interests with the means available to it.[47]

Norway does everything it can to destroy the Russian fishing industry. And that's good. That's how it should be. It's just a pity the Russian state isn't strong enough to defend the interests of its inhabitants in the same manner.[48]

Fourth, there is a clear tendency in the Russian press and among people involved in Russian fisheries to refer to the work of the Russian delegation to the Commission primarily in terms of defending Russia's national interests.

'There's a deep misunderstanding that Russia won this year's fishery negotiations in Murmansk,' says [deputy director] Vladimir Torokhov of Sevryba. To the newspaper Rybnaya Stolitsa in Murmansk [he] says that, quite the contrary, it was Norway that got its will. 'Our crafty neighbours had right from the start the same quota figures in their heads as the negotiations ended with. . . . We didn't learn the lessons of the previous negotiations and defended our national interests badly.'[49]

Of course, it's necessary to mention the argumentation of the Russian scientists in the negotiations [of the Joint Fisheries Commission], which, to a considerable extent, made possible the achievement of many acceptable results for the members of our northern fishery complex. . . . Our scientists really did a good job at the negotiations.[50]

[In the Joint Fisheries Commission, the Russian scientists] defended the Russian positions in a precise and well-prepared manner.[51]

Finally, the issue of Russia's national interests also figures in the debate concerning the internal organisation of Russian fisheries management and market access of the fish.

We need [a new regional fisheries department] which can defend us against Norwegian dominance in the fishing grounds. . . . We tried to get one of Sevryba's experts to assist us in [an ongoing trial in Norway], but it proved to be impossible. Sevryba as an organisation needs fundamental changes. It must become a strong department which not only co-ordinates us in the fishing industry, but also defends us [against the Norwegians].[52]

'If the cod quotas for next year are set in accordance with the scientific recommendations, we will deliver our fish to Canada and Portugal. Not a kilo will go to Norway.' The message from the fishing industry in Northwestern Russia was clear at a press conference in Murmansk yesterday evening. Norway was also accused of speculating in low quotas in order to gain control of the Barents Sea fish. . . . 'Norway stands firm on the low quotas because they calculate that Russian vessels will deliver their catches to Norway,' said [leader of the association of small shipowners in Northwestern Russia]. He was supported, among others, by the Vice Governor of Murmansk Oblast, Yuriy Myasnikov, reports news editor Andrey Privalikhin in Murmansk TV.[53]

In conclusion, representatives of the Russian fishing industry and management system appear to agree that the Norwegians should be seen as rivals, not as partners:

Fishery entrepreneurs and scientists of the Northern basin have held a 'round table' session in Murmansk to work out recommendations for [the] national delegation to future sessions of the Joint Russian–Norwegian Fisheries Commission. 'It's about time for us to think of the Norwegians as rivals rather than partners, in competition for marine bioresources,' said Gennady Stepakhno, head of the Northern Fishery Enterprises Association, at the opening of the meeting. Indeed, it seems to have been the general feeling of the meeting – along with the view that 'there is plenty of fish in the sea, and we must press our demands for an increase in the total allowed catch'.[54]

The Chernigov episode

The Fishery Protection Zone around Svalbard forms a particular jurisdictional and regulatory complex which in recent years has stirred dormant conflicts and exposed differing perspectives between Norwegian and Russian fishery authorities. The Svalbard archipelago is situated between the Norwegian mainland and the North Pole and was unclaimed territory up to the adoption of the Svalbard Treaty in 1920 and its coming into force in 1925.[55] The Treaty gave Norway sovereignty over the archipelago, but contains several limitations on Norway's right to exercise this jurisdiction.[56] Most importantly, all signatory powers enjoy equal rights to extract natural resources on Svalbard. While Norway claims that the Treaty refers only to the islands of Svalbard and their territorial waters, the other signatories assert that its non-discriminatory code applies also to the ocean area outside the territorial waters. Norway claims the right to establish an exclusive economic zone (EEZ) around Svalbard, but has so far refrained from doing so since this would clearly be unacceptable to the other signatories of the Svalbard Treaty. The Fishery Protection Zone around Svalbard, established in 1977, represents a 'middle course', aimed at ensuring a certain protection of the important feeding grounds for juvenile cod in these northern waters (see Figure 3.1). Separate quotas are generally not set for this zone, i.e. Russian and Norwegian fishers can fish here or in their own EEZ – the catches are in any event subtracted from their general Barents Sea quota.[57]

I have elsewhere argued that the Norwegian fisheries management in the

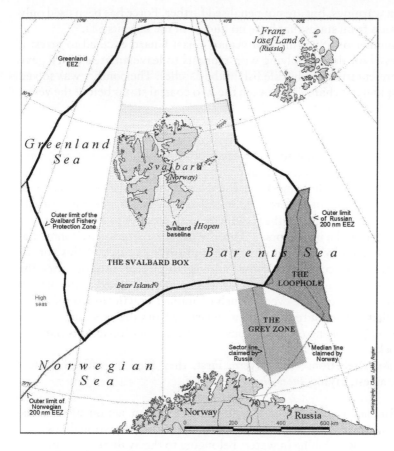

Figure 3.1 The Fishery Protection Zone around Svalbard.

Note: nm = nautical miles.

Svalbard Zone has been relatively successful despite the unclear jurisdictional status of this ocean area (Hønneland 1998b, 1999, 2000a, 2001). The bulk of the fishing vessels operating in the zone are Russian.[58] Although Russia has not formally accepted the Fishery Protection Zone, Russian fishers have largely complied with Norwegian fishery regulations in the area and also followed less formal requests by the Norwegian Coast Guard regarding where to fish in order to avoid excessive intermingling of juvenile fish. Russian vessels do not report to Norwegian fishery authorities about their catches in the zone, and Russian captains refuse to sign inspection forms presented to them by the Norwegian Coast Guard. However, they do welcome Norwegian inspectors on board, and the same inspection procedures are pursued in the Svalbard Zone as in the Norwegian EEZ.

To avoid provoking other states, Norway has chosen a 'gentle enforcement' regime in the zone, where violators of fishery regulations are given oral and written warnings, but are not penalised further. Force has been used only against vessels from states that have no quotas in the Barents Sea.[59]

In the summer of 1998, the Norwegian Coast Guard decided to arrest a Russian vessel for not complying with requests to leave an area which gave an overrepresentation of juvenile fish in the catches. The conflict was solved through diplomatic channels between the two coastal states before the vessel reached Norwegian port.

However, the existing order was seriously challenged when the Norwegian Coast Guard in April 2001 arrested the Russian trawler *Chernigov* in the Svalbard Zone. According to Norwegian authorities, the violations committed by *Chernigov* were exceptionally grave: an extra net had been attached to the trawl whose mesh measured less than half the legal size. Large amounts of juvenile fish were discovered on board at the time of the inspection,[60] and the crew cut the trawl wire in an attempt to avoid discovery of the extra net by the inspectors.[61] 'This is what we define as environmental crime, and that's why we give such a high priority to cases of this sort',[62] explained the chief constable of the Norwegian city of Tromsø after the vessel had been escorted to its port. This statement was typical of the official Norwegian reactions to the event, which contained practically no reference to the fact that the arrest represented a departure from the existing Svalbard Zone regime by focusing on the seriousness of the violation and its harm to the fish stocks.[63]

The Russian reactions were severe. First, they filed an official protest against the arrest, saying, *inter alia*:

> The Russian party cannot accept the Norwegian action against the Russian trawler as legal and in accordance with international law. . . . The trawler was fishing in waters belonging to the Svalbard archipelago, whose legal regime is regulated by the Svalbard Treaty of 1920, but outside the borders of the area where the named Treaty has application, that is in waters where the norms of international law on the high sea apply.[64]

Second, the Russian authorities decided to cease collaboration with Norway in the area of fisheries management shortly after the arrest. Among other things, the Russian delegation left an ongoing meeting in the Permanent Committee on Management and Enforcement Co-operation under the Joint Fisheries Commission.[65] Third, the arrest evoked extraordinarily strong reactions against Norway from Russian politicians at high levels. Most famous is the statement by Yevgeniy Nazdratenko, leader of the Russian State Committee for Fisheries, that Russian naval vessels should shoot at and sink Norwegian Coast Guard vessels in the Svalbard Zone and do nothing to save their crews.[66] The governor of Murmansk Oblast, Yuriy Yevdokimov, engaged in the debate, commenting on the *Chernigov* incident in the following way:

In the north, we are facing universal challenges related to the security of Russia. The Norwegians understand this, too. It is always like this: when one state is temporarily weakened, its neighbours will try to take advantage of it.[67]

Finally, the arrest of *Chernigov* served to bring to the surface what appeared to be much suppressed Russian frustration about what they perceived as a gradual expansion of Norwegian control over the Svalbard Zone fisheries. Partly, the discourse was linked to similar processes on land on Svalbard, notably the strengthening of Norwegian environmental regulations on the archipelago. The reports below show how this issue was framed both before and after the arrest of *Chernigov*.

There is still disagreement on the Russian–Norwegian fishery in the Svalbard Zone. According to the Paris Convention, signed more than sixty years ago,[68] our country has the right to conduct and expand industrial activity there. However, the Norwegians have recently introduced a range of measures aimed at pressing Russian vessels out of this area.[69]

On 5 June, the Norwegian parliament – the Storting – discussed a new bill on the regulation of environmental affairs on the archipelago of Svalbard. Oslo has also earlier been engaged in environmental protection in the Arctic: large parts of the tundra of Svalbard have been declared nature reserves and national parks, and parts of the ocean areas have been closed for fishing vessels. The new act makes the environmental regulations even stricter – to such an extent that our Ministry of Foreign Affairs found it to be in violation of the old Paris Treaty on Svalbard. More than that: on 6 June, Russia's Deputy Minister of Foreign Affairs, Aleksandr Avdeyev, stated that Norway was 'not far from wanting to establish nature reserves in places where Russian miners either extract coal or have plans to do so' and 'hence force us to leave the island.' ... After the arrest of the Russian trawler *Chernigov* in this zone, there is sufficient basis to speak about a Norwegian dominance in the zone and discrimination against Russian vessels.[70]

Russia wants extensive changes in the 200-mile fishery protection zone around Svalbard. 'Norway cannot continue to squeeze our fishers out of the area,' says leader of the catch department of the Murmansk-based fishing company AO Sevryba, Vladimir Torokhov. ... 'The leadership of Sevryba has continuously addressed our State Committee for Fisheries with demands that the Ministry of Foreign Affairs engage in the issue. It is high time the disagreements [between Russia and Norway] are solved.'[71]

Russia attempts to resume the country's surveillance of the fishery in the

Protection Zone around Svalbard. In August, the Russians sent one of their enforcement vessels to the area for the first time in twenty years. The enforcement vessel Strelets carried out a two-week inspection [trip] on the fishing grounds around Svalbard, informs Russian Fish Report. . . . The reason for the absence of Russian enforcement vessels in the area has been, and still is, lack of fuel. When the Russian [enforcement] vessel suddenly appeared on the fishing grounds, the reason was that the Russian company InterBarents had donated money to buy 300 tonnes of fuel. The Russian Border Guard has also appealed to other fishing companies to contribute fuel to enable the Border Guard to maintain its presence in the Svalbard Zone. On the Russian side, there is keen interest in demonstrating that one cannot accept total Norwegian domination in the area and in defending Russian fishers from 'sudden and breath-taking inspections' from the Norwegian Coast Guard, reports Murmansk newspaper *Rybnyy Murman*.[72]

Discourse on institutions

As the preceding discussion makes clear, there has been considerable frustration in recent years on both sides of the Norwegian–Russian border regarding the functioning of the existing management system of the Barents Sea fisheries. This section shows how this frustration has been expressed and enumerates the proposals for change that have been put forward from both sides.

The Norwegian discourse: institutions fit for sustainable management?

The official Norwegian view is that the existing management regime functions satisfactorily.[73] As expressed by the Norwegian Minister of Fisheries in autumn 2001: 'this is sustainable politics. The Labour Party . . . has consideration for both society and biology – in line with the precautionary principle'.[74] After a change in government, the new conservative Minister of Fisheries followed up a few months later: 'that next year's cod quota is above the recommendations from ICES does not mean that it's necessarily unsustainable. There is still no danger for the cod stock'.[75] The leader of the Norwegian delegation to the Joint Fisheries Commission the same year said after the two countries had signed the quota agreement for 2002: 'we have made sustainable decisions'.[76]

Criticism comes primarily from parts of the coastal fleet, environmental NGOs, regional political authorities and the political left.[77] Reference is made to Norway's international obligations according to international environmental agreements, in particular to the right of all relevant stake-holders to participate in the management process and to what is perceived to be an irresponsible exploitation of the Barents Sea fish stocks.

We are excluded [from the Joint Fisheries Commission] and hence denied participation in the debate about the future of our fisheries – just like the rest of the Norwegian population. . . . Unfortunately, Norwegian fishery policies are today carved out in a closed room, quite contrary to democratic rules of the game. It is only two weeks until [the Joint Fisheries Commission] meets in Murmansk in order, among other things, to establish total quotas for the Barents Sea for next year. We have no idea about the proposals Norway has put forward and hence no possibility to take part in the debate. . . . What kind of secrecy is it that we allow to develop in the Norwegian–Russian fisheries debate? A closed culture has developed in which both parties are caught in the negotiation game. The members of the Commission have, in a way, their own reality, with which the rest of the world cannot identify. The result is bad fisheries management.[78]

It's a cocoon, where the Directorate of Fisheries, marine science and the Ministry [of Fisheries] have created a strange fraternity. We are not informed about what is going on and whether it is in fact Russia that is pressing for higher quotas or if Norway is agreeing. The closed nature of the [Joint] Fisheries Commission violates democratic principles.[79]

If Norway had lost the same sums on ill-judged political decisions related to a small piece of road in Eastern Norway as we lose every day in the Barents Sea, the cries for a board of inquiry and a public hearing would be deafening. But when it comes to a fishery policy that has so significantly bereaved us of years of solid income, and which threatens the habitation of an entire region of the country, no demands for an inquiry are heard. The [fishery] policy throughout the 1990s has been an ill-judged scandal. It would be quite unbelievable if no one would have to answer for it and measures were not taken to avoid the same thing happening again in the future. . . . Much indicates that a [parliamentary] hearing would be exceptionally unpleasant and awkward for the politicians, civil servants, scientists and prominent representatives of the fisheries organisations that would have to be summoned.[80]

The cod has become too important to be left to civil servants and players within the fishing industry. The [Joint] Norwegian–Russian [Fisheries] Commission is dangerous because, in a most irresponsible manner, it has allowed the quotas to skyrocket. It has become totally superfluous. The establishment of quotas should be left completely to independent scientists.[81]

[We demand] that the decision-making system for [fisheries] management between Norway and Russia opens for a broader participation of stakeholders along the coast and that the Storting [Norwegian parliament] becomes more involved in Norwegian quota politics.[82]

The Storting [Norwegian parliament] has never been informed of the Norwegian position in the Norwegian–Russian fishery negotiations. A parallel case would be not informing the Storting about the speed of the extraction of oil on the shelf or the Norwegian position in the international climate negotiations. The same power structure that decides the quota politics is now set to define how the precautionary principle is to be applied in the management of cod. In this connection, the authorities indicate that they want to reduce the possibilities of marine scientists to express whether different quota levels are in accordance with the precautionary principle. The county councils of Finnmark and Nordland recently demanded participation for the counties in the Norwegian delegation to the fishery negotiations [with Russia]. This was refused by the Ministry of Fisheries. . . . It is probably only the Storting that can elevate the issue to a political level that makes it possible to adopt a new policy. It is Norwegian environmental politics, the fate of hundreds of fishery communities and large export incomes that are at stake.[83]

In late 2001, the Norwegian parliament – the Storting – responded to both explicit and implicit complaints by critics of official fisheries management. All the political parties on the Select Committee on Trade and Industry stated that Norwegian fishery policy was not sustainable:

The Committee notes with concern that the agreements that have been concluded between Norway and Russia in recent years have not given sufficient consideration to a sustainable management of the cod stock. . . . We note that the negotiation result [for 2002] is at a level far above scientific recommendations and emphasise that the main principle for future negotiations is to give priority to the restoration of the stock.[84]

In sum, there is a growing public concern in Norway about the establishment of cod quotas in the Joint Fisheries Commission since the late 1990s. There is an increasing demand that what has so far primarily been the domain of the Ministry of Fisheries, is opened up for public scrutiny and participation by other stakeholders than those representing directly involved groups within the fishing industry. Environmental NGOs and regional authorities in Northern Norway are the most insistent in this respect. Further, there are signs that Norway's international fishery relations, which for years have been mainly a bureaucratic matter enjoying very limited parliamentarian interest, are again being elevated to the level of active parliamentary politics.[85] Hence, the Joint Fisheries Commission is seen as not necessarily fit for conducting the kind of fishery politics that has been defined by the highest political authorities in Norway. Interestingly, the issue of 'sustainability' is generally given as a rationalisation of the critique of current management practice; we will return to this later.

The Russian discourse: institutions fit for securing national interests?

To the extent that discussions about the existing management practice of the Barents Sea fisheries take place in Russia, Russia's national interests will be at issue. There is a growing concern about whether the Joint Fisheries Commission is in fact fit to securing Russian national interests. As shown by the discourse on interest above, the attitude of Russian fisheries actors is that Norwegians take advantage of Russia's unfortunate economic and political situation. As expressed by a former Soviet Deputy Minister of Fisheries and current advisor to the country's fishery authorities at both federal and regional (northwestern) level:

> Diplomatic relations with Norway within the fisheries sector have, mildly speaking, been faltering for several years. . . . It all began a couple of years ago with what seemed like a trifle: because of our economic difficulties, we were not able to carry out several joint scientific programmes. Then concessions were made [from our side] in a range of negotiations. Although they might appear insignificant at first glance, they had unfortunate consequences. For instance, we too early agreed to introduce selection grids in the cod trawl according to the Norwegian scheme. . . . Or take the last negotiations on a trilateral agreement between Iceland, Norway and Russia. Yes, this agreement was necessary; no one disagrees on that fact. But it should not have given special rights to the party that is not a co-owner of the Barents Sea resources. Today, our fishers quite rightly raise the question about the Norwegian inspections in the Svalbard area, which are very rigid – sometimes outspokenly biased. For me, who knows the Norwegian fishery control system well, this sometimes seems strange. What worries me, is that their inspectors too frequently close fishing grounds with doubtful motives – alas, there is too much undersized fish in the catches. . . . There has been a generation shift in our fishery diplomacy. Some have retired because of old age. Others have just 'retired', and, as a result, the logic of our management system has broken down. But that being as it is, does it give the other party the right to take advantage of his partner's mistakes and grab for himself more than is righteously his?[86]

Correspondingly, Russian trust in ICES seems to have shrunk in recent years.

> Central participants in the Russian delegation [to the Joint Fisheries Commission] show no mercy to the International Council for the Exploration of the Sea, ICES. While [the Joint Fisheries Commission] will probably set a cod quota for next year that lies far above the scientific recommendations and not take into consideration the scientists' views on [sustainable] fishing patterns, the powerful shipowner and [State] duma representative Vladimir Gusenkov states that ICES is far

from having the necessary credibility and objectivity. 'ICES reminds me a bit of the scientific advice that we had during the Soviet era, where all recommendations were in accordance with the government's special interests. The difference is that ICES defends Norwegian interests in each and every way.'[87]

Defining major discourses

What are the main discourses on marine living resources in the Barents Sea? The current section sums up the main findings of our discussion so far in what I have chosen to call the 'sustainability discourse', the 'Cold Peace discourse', the 'seafaring community discourse' and the 'pity-the-Russians discourse'. For each discourse, its basic assumptions, the basic entities forming it, its main discursive coalitions and story lines are defined. The interrelation between the discourses will be dealt with in the last section of the chapter. While this section mainly organises already presented arguments into the four discourses, new information is also added, not least when discourses on marine living resources can be said to form part of more overarching discourses in society. The story lines that are presented as typical of each discourse are direct quotes from or slightly paraphrased versions of the interview material and media extracts already mentioned. I do not claim that these concrete story lines have had the tangible and immediate effects on their respective discourses that Hajer (1995) (see Chapter 1) asserts that story lines often do. Rather, I believe they capture central aspects of the discourse regardless of whether this was the first time they were uttered or not. Hence, similar utterances are believed to have had an effect or given direction to the discourse in question when they were first produced although their authors are unable to point at the exact time at which this took place.

The 'sustainability discourse'

The Norwegian discourse on the management of the Barents Sea fisheries has above all evolved around the issue of 'sustainability': has the establishment of quotas been sustainable or in accordance with the precautionary principle? Within this more general discourse, two 'discourse coalitions' are discernible, sharing some basic assumptions but diverging as far as the prescribed solutions are concerned: the 'official' and the 'critical' sustainability discourse. Both variants of the 'sustainability discourse' maintain that disputes about quota settlement are mainly disputes about sustainability. This is what the debate is about: should the cod quota be set at 200,000 tonnes or 500,000 tonnes or at any other proposed level? Which alternative is most sustainable and sufficiently precautionary? Discussions about interest and institution are seemingly subordinate to this overarching question: are current management practices 'sustainable' or 'unsustainable'? Solutions related to whose interests should be favoured or which institutional arrangements are most suitable follow from the answer to the question concerning whether the observed management practice is sustainable or not. The 'official' and the 'critical' sustainability discourse also concur in their main objects of discussion: both focus on the future of fishers' households and coastal communities as the main human (or 'non-fish') objectives of a sustainable fisheries management.

The two variants of the 'sustainability discourse' diverge when it comes to how the above questions should be answered. The 'official' variant says that quota settlements have been sustainable; one has only chosen to build up the cod stock at a slightly lower pace than that proposed by ICES scientists. On the other hand, the 'critical' version of the discourse says that management practices in the 1990s have brought the Northeast Atlantic cod stock to the verge of extinction. It might have been 'necessary' to depart from the scientific recommendations in the quota settlements, e.g. in order to maintain the bilateral regime with Russia, but management practices have definitely not been sustainable.

The discourse coalition forming the 'official' sustainability discourse is made up of the ruling political parties, civil servants and the main fishers' organisations in the management process, that is, the Norwegian Fishers' Association and the Norwegian Association of Shipowners.[88] Although occasionally indulging in cautious criticism of the tendency to set quotas far above the scientific recommendations – blaming the Russian party for the 'necessity' to do this – this discourse coalition by and large maintains that last years' quotas do not represent any significant harm to the Northeast Arctic cod stock. Politicians and civil servants tend to emphasise how actual management practice accords with internationally acclaimed principles of sustainability and precaution – the rebuilding of the cod stock is taking only a bit longer than the ICES opinion would like to see. Fishers and their representatives tend more to question the 'correctness' of the recommendations coming out of ICES and claim that it would not have been 'right' to be as cautious as ICES recommends.

The 'critical' discourse coalition, on the other hand, maintains that quota settlements have not met the criteria of sustainable and precautionary fisheries management. The discourse is composed of representatives of the coastal fishers – mainly those organised in the 'alternative' Norwegian Association of Coastal Fishers, the only fraction of Norwegian fishers that has chosen to stand outside the Norwegian Fishers' Association – environmental NGOs, parties on the political left and to some extent also scientists.

The most conspicuous trait of the 'sustainability discourse' is, however, its tendency to answer all questions – whether related to issues of interests or institutions – in terms of sustainability or non-sustainability. Representatives of the 'official' discourse coalition could in principle have claimed that the relatively high cod quotas in recent years have saved the Norwegian fishing industry from difficult distribution battles between various fleet groups at the national level and therefore represented good political handiwork. But it would not have been as 'politically correct' as maintaining that the quotas have been 'sustainable', thus, in a way, avoiding the more traditional interest-based explanation. (I shall return to this issue in the concluding section of the chapter.) Likewise, adherents of the 'critical' discourse coalition could simply have stated that they do not like trawlers and range the vitality of small coastal communities above the opportunities for large shipowners to accumulate even more capital. Reference to widely acclaimed principles of sustainability and precaution, however, adds greater momentum to their arguments. For instance, when Norwegian authorities in April 2001 chose to break with two decades of established practice and decided to arrest a violator in the Fishery Protection Zone around Svalbard, the Russians, overly critical of and not so little surprised at this change in direction, were met by a unanimous 'sustainability chorus' in Norway: the arrest was necessary to ensure sustainable fisheries management. What this 'chorus' did not mention was that similar violations, no less dangerous to the area's fish stocks, had been committed more or less regularly for decades, without provoking graver responses from the Norwegian Coast Guard than written warnings. Hence, it could be said that 'sustainability' seems to have become a sort of mantra for Norwegian fishery authorities when seeking to justify 'difficult questions' in the country's fishery politics *vis-à-vis* neighbouring states (see Table 3.3). We will return to this topic too in the concluding section.

The 'Cold Peace discourse'

Whereas the Norwegian 'sustainability discourse' focuses on the conflicting interests between various sub-groups of society, the Russian discourse on the establishment of TACs for the Barents Sea cod is strictly state-centred. Where

Table 3.3 The main assumption, entities, discourse coalitions and story lines of the 'sustainability discourse'

	'Official'	'Critical'
Basic assumption	• disputes about quota settlement are disputes of sustainability	
Basic entities	• fishers' households, local communities	
Discourse coalitions	• politicians in office, bureaucrats, Norwegian Fishers' Association, Norwegian Association of Shipowners	• Norwegian Association of Coastal Fishers, environmental NGOs, scientists, the political left
Major story lines	• 'Current fisheries management is in accordance with Norway's international obligations.'	• 'The closed nature of the Joint Fisheries Commission violates democratic principles.' • 'This is gambling with the resource base, and we should oppose it with everything we have of expertise and weight.'

the former speaks of short- versus long-term interests of different types of vessels and regions, the latter views quota settlements – quota settlements that by definition are divided on a 50–50 basis between Norway and Russia – as a battle of national importance between the two states involved. First, the two states are the basic units in the discourse. Matters of TAC establishment are not seen as matters of distribution between different groups in the countries; this is the prerogative of the 'sustainability discourse'. Second, both the 'Russian' and the 'Norwegian' interests in the issue are seen as stable and unitary, rather than fluctuating and multifaceted. Third, the 'opponent', i.e. Norway, is seen as a strictly rational actor, capable of calculating the value of its own interests precisely and minutely (because it knows exactly what its best interests are). Fourth, the interests of the two states are considered to be mutually opposed and the establishment of TACs a zero-sum game: one party can gain only what the other loses. Consequently, states are believed to be incessantly seeking to destroy other states in never-ending attempts to maximise their own interests. Compare some of the statements referred to earlier: 'It is always like this: when one state is temporarily weakened, its neighbours will try to take advantage'; 'There is nothing special about this – every country defends its own interest with the means available to it.' Or, as expressed by a Russian fisheries researcher, pinpointing the Russian perception of the quota establishment exercise as a zero-sum game: 'Of course, it's in Norway's interest to ruin Russia; this is simple economic theory.'[89]

The Russian discourses on marine living resources in the Barents Sea during the 1990s form part of a general Russian discourse on the country's relations with the West: what I will here label the 'Cold Peace discourse'. The term refers to the mounting sense of disappointment in the West felt by many Russians from the early and mid-1990s, as it became increasingly clear that the political and economic reforms were not bringing the results many had hoped for. Combined with a sense of resentment against NATO expansion eastwards, and the 1999 NATO intervention in Kosovo, many Russians became convinced that the co-operative attitude of the West at the end of the 1980s and the beginning of the 1990s was false, a way to press reforms on Russia that it was known would not work in the Russian setting in any event. The motive of Western powers was allegedly to weaken Russia even further while, at the same time, taking the advantage to boost own military and economic power. The process was conducted under 'positive' slogans such as 'democratisation' and 'introduction of market reforms', but many Russians came to see it as little more than a continuation of the Cold War East–West struggle, a situation that gave rise to the term 'Cold Peace'.[90] By the end of the decade, many had become disillusioned with the West and instead sought answers in 'traditional' patriotic values. As expressed by the leader of the Murmansk regional duma commission for patriotic upbringing in 1999:

After some years of liberal reform . . ., the concept [of patriotism] is again attracting lively interest. To an increasing extent, the columns of patriots are joined by those who only recently were ignoring the national interests of the state, and instead tried to convince us of the priority of certain abstract universal human values.[91]

In Northwestern Russia, growing distrust in the West was given further sustenance by disappointment in the results of the Barents Euro-Arctic regional co-operation venture (see Chapter 2). The Nordic countries had held out expectations of large investments in Northwestern Russia when they launched this regional initiative in 1993, expectations it proved very hard to live up to. As I observed in a 1998 article on identities in the Barents Euro-Arctic Region:

> A further indication of new contrasting between East and West [in the Barents Euro-Arctic Region] is the sense of bitterness with their Nordic neighbours observable among Russian politicians, civil servants and businessmen recently. This seems to be the result of disillusionment regarding the achievements of the Barents co-operation so far. During several visits to Murmansk in the course of 1997, conducting interviews with the mentioned categories of people, the author of this article was repeatedly instructed to go back to Norway and report how disappointed they are. The message is as follows: 'you have done nothing to help us, so don't send more delegations! We are sick and tired of your endless talk. Please, leave us alone!'
>
> (Hønneland 1998c: 292–293)

Add to this the resentment in Northwestern Russia towards the fact that most of the Barents Sea cod caught by Russian vessels has since the early 1990s been delivered in Norway and other Western countries. This has been a disaster for the land-based fish-processing industry in Murmansk, once the largest fish industry in the entire Soviet Union. Although it is Russian captains on fishing vessels who choose to deliver their catches abroad – the main reasons being proximity to fishing grounds and less bureaucratic delivery procedures than in Murmansk – most Russians interpret the development as 'Norwegians stealing our fish'.

Hence, the events given most attention in this chapter – the establishment of cod quotas at the turn of the millennium and the *Chernigov* episode in 2001 – took place in an environment characterised by increasing Russian suspicion regarding the intentions of Western countries in their dealings with Russia. When Norway 'demanded' a drastic decrease in cod quotas based on scientific opinion, which the Russian party did not find altogether legitimate – among other actions, Russia filed an official protest against the level of reference points established by ICES – it was seen as yet another attempt by a Western neighbour to take advantage of Russia's floundering economic situation. When the Norwegian Coast Guard for the first time arrested and penalised a Russian fishing vessel in the Svalbard Zone, it was taken as confirmation of Norway's intention to 'squeeze our fishers out of the area' and 'force us to leave the island'. The Norwegian chants of 'sustainability, sustainability' made little sense to the Russians, fuelling instead Russian suspicions that its Western neighbours were up to no good with their hidden agenda. The 'Cold Peace discourse' provided an arena for speculation: did the Norwegians want to reduce the quota in order to keep cod prices high on the world market, or was their intention to 'ruin Russia'? Story lines such as 'Every country defends its own interests with the means available to it', 'Norway does everything it can to destroy the Russian fishing industry, and that's how it should be' and 'Of course, it's in Norway's interest to ruin Russia; this is simple economic theory' provided ready explanations to complex and otherwise incomprehensible situations (see Table 3.4).

The 'seafaring community discourse'

We have seen evidence for considerable scepticism among fishers on both sides of the border towards the stock assessments made by marine scientists. A common theme in this scepticism is directed at scientists' apparent inability to assess stocks correctly because of their 'absence from the sea'. According to this view, fishers are in a better position to have a qualified opinion about the situation in the sea since they are observing it all the time. The scientists, for their part, spend only part of their time at sea, sitting otherwise in their comfortable offices in large cities, engaging in academic calculations on the size of the stocks: 'Everyone who works at sea can see what's going on and that the scientists are wrong'; 'Contrary to the scientists, who go out at sea a couple of times a year, we follow activities at sea every day throughout the year – we know what we are doing'. There is the further perception that whatever the scientists say, the actual situation is the opposite: 'There have never been such large amounts of accessible fish of exactly those species that the scientists can hardly find at all'; 'Never have the cod been so easy to catch, and never were there such large amounts of them'. At the same time as the

Table 3.4 The main assumptions, entities, discourse coalition and story lines of the 'Cold Peace discourse'

Basic assumptions	• states are invariably in competition with each other • relations between states are zero-sum games
Basic entities	• states
Discourse coalition	• Russian politicians, bureaucrats and shipowners; the Russian press and public
Major story lines	• 'Every country defends its own interest with the means available to it.' • 'Norway does everything it can to destroy the Russian fishing industry, and that's how it should be.' • 'Of course, it's in Norway's interest to ruin Russia; this is simple economic theory.'

scientists see a severe crisis looming, fishers see 'bumps in the ocean of fish of all sizes' and 'so much saithe that we can't manage to keep the trawl net away from them'.

As I have suggested, such comments are more a reflection of traditional distrust of scientific experts among fishers, and more generally, of urban experts set to tell them about conditions at sea. In my discussion on compliance in the Barents Sea fisheries (Hønneland 1998b, 1999, 2000a, 2001), I observed that, contrary to what many would expect, both Russian and Norwegian fishers had a very high degree of trust, even a sense of 'comradeship', with the Coast Guard inspectors. This sense of comradeship can be illustrated by the following interview extracts: 'It's impossible to be a fisher and at the same time be against the Coast Guard' (Norwegian fisher); 'We have a good relationship, the inspectors are polite, we respect each other here at sea, we all have a hard job' (Russian fisher). There appears to be a mutual respect among 'those at sea', a sense of comradeship across functional roles based on traditional maritime values among fishers and others who have their occupation at sea. In contrast to the scientists, who only spend a week or two at sea every now and then, the Coast Guard inspectors work under the same conditions as the fishers themselves. What is more, they are not 'academic wise guys' claiming to have knowledge about the fish stocks that the fishers do not have. They all belong to the same 'seafaring community'.[92] Expressions of a common sense of fellowship are never far off when distrust in scientific estimations and scepticism towards urbane scientific experts are voiced (see Table 3.5).

The 'pity-the-Russians discourse'

As mentioned earlier in this chapter, the gap between the Russian and Norwegian positions on the cod TAC in the Joint Fisheries Commission became evident for the first time in 1999, when scientists recommended a drastic reduction in the quota. The negotiations broke off for several days and talks were not resumed until the evening before the parties were to leave for home. The Norwegian delegation agreed to set a quota that was far closer to the Russian position than the Norwegian one on condition that the

Table 3.5 The main assumption, entities, discourse coalition and story lines of the 'seafaring community discourse'

Basic assumption	• those at sea know best the situation at sea
Basic entities	• individuals
Discourse coalition	• Russian and Norwegian fishers
Major story lines	• 'Contrary to the scientists, who go out at sea a couple of times a year, we follow activities at sea every day throughout the year – we know what we are doing.'
	• 'While scientists speak about a crisis, the fishers experience the best catches ever.'

following statement – extraordinary in such a context[93] – be included in the protocol from the session:

> The Norwegian party notes that the level of the cod quota is alarmingly high in consideration of available stock assessments and the recommendations of ICES. Taking into account the difficult conditions of the population of Northwestern Russia ..., Norway has nevertheless found it possible to enter into this agreement.
>
> (Ministry of Fisheries 1999: 2)

The key words in this statement are 'the difficult conditions of the population of Northwestern Russia'. The message is that Norway feels 'compelled' to go in for a quota far above the scientific recommendations for humanitarian purposes; a reduction in the quota would have meant too heavy a burden on an already distressed population. We find replicated here the predominating Norwegian discourse on Northwestern Russia during the 1990s, what I have chosen to call the 'pity-the-Russians discourse'. The essence of this discourse is that the population of Northwestern Russia is suffering from deep poverty and is in constant need of help from its wealthy neighbours in the West. In the early 1990s, the environmental problems of Northwestern Russia – the danger of nuclear radiation and the results of air pollution (see Chapters 4 and 5) – were at the heart of Norwegian media interest and political initiatives towards Russia. The social problems of the area claimed greater attention from the mid-1990s. The portrayal of Northwest Russians as 'needy' culminated during the economic crisis in Russia in autumn 1998. Hordes of Nordic journalists descended on the Kola Peninsula in search of 'disaster stories'. In the event, they found only a few old people in the countryside – Murmansk Oblast is one of the most heavily urbanised regions of Russia – who could give substance to their stories of people 'on the verge of starvation'.[94]

A particularly interesting point is the difference in which 'the crisis in Northwestern Russia' was perceived in Norway and Russia. The Norwegians designated it a catastrophe and called for immediate action. In Murmansk, however, the Scandinavian humanitarian aid had been given a jaundiced welcome for a long time by large segments of the population – such aid had been forthcoming long before the 1998 crisis. For most Russians, the mere fact of humanitarian aid coming from the outside is enormously humiliating. In the Murmansk regional administration, there were complaints about the quality of some of the donated commodities.[95] Medicines and food had passed their sell-by dates; clothes and shoes were worn out and useless.[96] Further, Norwegian concerns about hunger in Murmansk Oblast and plans to accept 50,000 refugees were given short thrift. This is what a journalist in *Polyarnaya Pravda* wrote of a trip she made to Norway at the time:

> The well-worn phrase that 'we feel sorry for Russia' comes automatically, in particular when you're assaulted with questions from those

worried Norwegians. Every Russian-speaking person is apparently to be interrogated: is it true that there is hunger in your country? And then they go on to say that they have built refugee camps on the border for 50,000 people. Assurances that there is neither hunger nor prospects of any mass emigration do not deter the foreigners from sending us humanitarian aid.[97]

For most Russians, the idea of migrating to Norway sounds as ridiculous as it is incomprehensible. It should be mentioned, however, that Norwegian plans to house 50,000 Russians had been prepared several years ahead with a nuclear incident in the area in mind; this aspect was initially not mentioned either by the Norwegian or Russian media. In addition, however, Governor Yevdokimov was clearly engaged in some kind of duplicity: when in the Nordic countries, he humbly asked for humanitarian aid, describing the situation as catastrophic. At home, however, he said that Murmansk Oblast could cope on its own, implying thereby that foreigners who wanted to lend a hand were pretty naive:

> There is no tragedy, there is no catastrophe in our region. There is no reason to expect 50,000 refugees on Norwegian territory. We can cope without their humanitarian aid. It only helps us to cope more quickly.[98]

The image of starving and impoverished Northwest Russians continues to be a staple of Norwegian media and politics. At the time of writing, the Norwegian Red Cross is running a television campaign to raise funds for soup stations in Northwestern Russia. 'For only [a very small sum of money] a day, you can give a child in Northwestern Russia a portion of soup,' announces Thorvald Stoltenberg on the television. Stoltenberg was Minister of Foreign Affairs when the Barents Euro-Arctic Region was established, and is currently president of the Norwegian Red Cross. He might be perfectly right in asserting that a meal of soup can be purchased for the given sum of money and that the Norwegian Red Cross has the infrastructure to serve this soup to children in Northwestern Russia. It is probably also correct that there are children in Murmansk who experience hunger, but the same can also be said about Oslo, London or New York. The point is that the Red Cross television campaign serves to reproduce a discourse which says that starvation is a normal thing for little children in Murmansk. In connection with an evaluation of media and competence projects financed by the Barents Euro-Arctic Region Programme (Jørgensen and Hønneland 2002), I interviewed the leader of the Barents Press office in Murmansk, the aim of which is to further co-operation between journalists from the East and West in the Barents Euro-Arctic Region. She complained that Nordic journalists are interested only in writing 'disaster stories' about Northwestern Russia, and mentioned the *Kursk* incident, environmental degradation and social problems.[99] On occasion, she had succeeded in attracting the interest of foreign journalists to other types of stories, but their editors back home cut the stories because they were not 'saleable'. Put differently: those types of reports did not interlock with the prevailing discourse and might therefore not have been 'understood' by the Norwegian public! What the editors needed, was what people expected from Northwestern Russia, i.e. tragedies and calamities.

Hence, the decision of the Norwegian delegation to the Joint Fisheries Commission in 1999 to comply with Russian demands not to reduce the quota significantly took place at a time when the impression of Northwestern Russia to be gained from the Norwegian media and politics was one of impoverishment and starvation. While I will not in this context speculate further as to the motives of the Norwegian delegation – obviously, it had to choose between two evils[100] – it can be argued that the 'pity-the-Russians discourse' in the Norway society provided a 'pretext' for setting aside scientific opinion: 'Sorry, but the lives of small children are more important than sustainability of the cod stock!' However, apart from the fact that the image of an impoverished population in Northwestern Russia is exaggerated beyond recognition, the 'increase' (or actually: the non-reduction) in the cod quota has not measurably benefited people in the area either. First, Murmansk Oblast is no social 'disaster area'; it is one of the richest regions of the Russian North and one of the few subjects of the entire federation that is a net donor to the state budget. Second, the Barents Sea cod is one of the natural resources in the area from which society receives least revenue since it is mainly delivered abroad. An 'increase' in the cod quota mainly serves to benefit the shipowners in Murmansk, i.e. to make the richest layer of the population even richer. The image of the 'starving children' went right home with the Norwegian public, and was probably also a 'reality' from the point of view of the Norwegian negotiators! The story line saying that 'the difficult circumstances of the Northwest Russian people necessitate a high cod quota' renders possibly unsustainable management practices more defensible. The 'pity-the-Russians discourse' makes an otherwise unacceptable political solution legitimate. Language in use influences politics (see Table 3.6).

Discourse genealogy and power

The Norwegian discourse on fisheries management in the Barents Sea has been a very vocal rehearsal of the 'sustainability chorus'. Whatever the issue,

Table 3.6 The main entities, discourse coalition and story line of the 'pity-the-Russians discourse'

Basic entities	• individuals, sub-groups in society
Discourse coalition	• the Norwegian press and public, Norwegian politicians, bureaucrats and NGOs
Major story line	• 'The difficult circumstances of the Northwest Russian people necessitate a high cod quota.'

i.e. disregard for scientific opinion or stricter Norwegian enforcement in the Svalbard Zone, excuses have calmed ruffled feathers with the refrain 'it's still within sustainable limits' and explanations evoked 'the need for sustainability'. When the Norwegians started arguing in 1999 in the Joint Fisheries Commission that quotas must be cut, and that the issue was one of sustainability, Russian suspicion was aroused. The Russians generally think that cod stock indications are not really too bad, and that ICES reference points are consequently too high. The establishment of cod TACs is certainly *not* an issue of sustainability in their eyes, which is why they nurture suspicions of a hidden agenda behind the Norwegian stance. The dominant 'Cold Peace discourse' in Russia provides an arena for speculation. Seen through the lens of this discourse, states are ceaselessly at loggerheads over material resources, and it is always in one state's interest to damage the interests of another (even if without seeing any immediate gains oneself). In addition, the Russians have 'observed' that the West has been consciously trying throughout the 1990s to ruin Russia under the cover of 'democratisation' and 'economic restructuring'. The Norwegian 'sustainability chorus' is easily recognisable as a foil for defending Western national interests, whether in the form of maintaining high world market prices for cod or simply damaging Russia as a competitor. Story lines such as 'Every country defends its own interests with the means available to it' and 'Norway does everything it can to destroy the Russian fishing industry' give sense to otherwise unclear motivations.

At the same time, the 'seafaring community discourse' feeds on distrust on both sides of the border of the scientists' pessimistic prognoses for the Barents Sea cod stock. The discourse portrays the science as fundamentally out of step with the experiences of the men at sea, fishers and other types of sea folk. Scientific language is viewed with disdain, and the 'seafaring community discourse' grips at any opportunity to exaggerate claims that the science is wrong. Its main objective is to underpin the notion that fishing should be left to the fishers; experts are not needed. Its most important consequence is to weaken the view from science that the Barents Sea cod stock is in danger of overfishing by inflating the margins of uncertainty that always accompany scientific prognoses. Its purpose, in other words, is to weaken the arguments of the 'sustainability discourse' and strengthen the conclusions (if not the arguments) of the 'Cold Peace discourse'.

As I write, divergent management views seem to have reached deadlock: the Norwegian 'sustainability discourse' and the Russian 'Cold Peace discourse' are based on incompatible premises. The 'seafaring community discourse' supports Russian argumentation, reducing, as it does so, the abilities of the Norwegians to win support. Then the 'pity-the-Russians discourse' offers a way out of the deadlock: the Norwegians are ready to give in on (unreal but probably sincerely felt) humanitarians grounds; the 'difficult circumstances of the Northwest Russian people' means that all good sustainability intentions must be dropped. The result is a type of management that, by definition, infringes the internationally recognised precautionary approach.

So who is the winner in the fisheries management battle over the Barents Sea fish stocks? Norway? Russia? The ocean-going trawlers or the Norwegian coastal boats? The fishing industry or environmental conservationists? The issue of power is not at the heart of this book's discussion. Its aim is to show how language practices *influence* politics, not to provide a *complete* answer to all questions related to redistributive outcomes and power relations. Some few remarks in that area are nevertheless not out of place. At first glance, it might seem as if Russia 'won' the quota battle during the period 1999–2001: the Barents Sea cod TACs were set at levels far closer to the Russian than to the Norwegian position. Likewise, it seems as if the arguments purveyed by the Norwegian ocean-going fleet have prevailed over those of the country's coastal fleet and environmentalists. These might be correct conclusions, and the outcomes could possibly be explained by other approaches than discourse analysis (such as more traditional interest or power-related approaches). The point in this connection is that prevailing discourses in society provided 'windows of opportunity' for the given outcomes: the Russian 'Cold Peace discourse' made it possible for wealthy shipowners in Murmansk – and Russian fisheries civil servants who, according to popular belief, get their piece of the quota revenue cake – to assert that Norway's wish to reduce the quotas was nothing else than yet another Western attempt to take advantage of Russia's disrupted economic situation. The 'seafaring community discourse', prevalent among fishers in both Norway and Russia, proved a useful assistant in this context. The Norwegian 'pity-the-Russians discourse' showed a way out of the deadlock, at the same time securing the interests of the Norwegian fishing industry and relieving Norwegian politicians and civil servants of difficult redistribution tasks. The Norwegian 'sustainability discourse' seems pretty futile in the quota settlement context, but it has probably secured Norwegian power interests in questions related to enforcement in the Svalbard Zone (compare the *Chernigov* episode). After more than two decades of cautious and rather sensitive management of the zone, Norwegian enforcement policy took on a bolder face at the turn of the decade. The Norwegian – and more generally, Western – 'sustainability discourse' provided the arguments necessary to ensure at least internal support for this bolder approach and reduce Russia's chances to gain external support in its dispute with Norway.

4 Discourses on nuclear safety

It wouldn't have been a problem for us [in Norway] if *Lepse* had sunk in Murmansk harbour. This is a local problem, but *Lepse* became a focal case for Norway.

(Norwegian civil servant)

Lepse was given as top priority [by the Norwegian side], but it's become a great failure. If we hadn't docked *Lepse*, it might have sunk. I don't understand why Norway pursued this course. I don't understand how you can sit there pushing pens while nothing is being done about the ship. Do you really want to take that responsibility?

(Russian civil servant)

Introduction

Radioactive pollution in Russia and Eastern and Central Europe has been designated one of the major environmental and security policy challenges in the European Arctic region in the post-Cold War period. There is widespread nuclear activity in the area, both civilian and military, particularly in Northwestern Russia. Hazards stem from unsatisfactory storage of large quantities of radioactive waste, decommissioned nuclear submarines awaiting dismantling, and the continued operation of unsafe nuclear power plants. As seen in Chapter 2, extensive Western efforts, in particular from Norway and the USA, have aimed to reduce the threat of the potential spread of radioactive pollution from Northwestern Russia since the early 1990s. The main question in this chapter concerns how Russian and Western discourses have contributed to shaping perceptions of nuclear safety problems, as well as concrete endeavours directed at their solution, in various international regimes. Empirically, the chapter draws on my participation in the evaluation of the Norwegian Plan of Action for the Implementation of Report no. 34 (1993–1994) to the Storting on Nuclear Activities and Chemical Weapons in Areas Adjacent to our Northern Borders (Ministry of Foreign Affairs 1995), hereafter referred to as the Plan of Action (Hønneland and Moe 2000). This evaluation involved interviews by the author with a range of participants in activities financed over the Plan of Action on both the Russian and the Norwegian side.

The chapter is divided into three main sections. First, we draw a picture of the main features of the evaluation of the Plan of Action, covering co-operation at state level between Russia and Norway and the implementation of a selection of concrete joint projects. Second, the major discourses on nuclear safety are defined. This builds on the evaluation material but places it in a wider context. Third, the discussion is rounded off with some notes on discourse genealogy and power. Given the complexity of nuclear safety management in Russia, a brief overview of the Russian regulatory system might be warranted before we embark on the main discussion. This presentation also builds on material from the evaluation.

Nuclear safety management in Russia

The main governmental body in Russia on nuclear safety issues is the Ministry of Atomic Energy (Minatom). Minatom was established in 1986 as a Soviet ministry and merged in 1989 with the powerful Ministry of Medium Machine-Building, responsible for development and production of nuclear weapons and reactors in the Soviet Union. Minatom was assigned responsibility for all aspects – civilian and military – of the nuclear energy industry and had about a million employees within its structure. This set-up continued when Minatom was reorganised as a Russian ministry in 1992. In 1998, responsibility for nuclear waste from military establishments was also transferred to Minatom. Minatom is now the main agency representing the Russian Federation in bilateral and multilateral discussions on atomic installations and nuclear waste and also oversees the implementation of joint projects with other states. The concrete implementation of such projects is, however, largely delegated to subsidiary or associated bodies, the most important of which, in the co-operation with Norway, is the so-called Interbranch Co-ordination Centre Nuklid.

Nuklid was established in 1990 to co-ordinate attempts at commercialisation within the nuclear sector of the Soviet Union. It is not formally part of the Minatom structure , but is part of the Minatom 'system', and is organised as a so-called unitary state enterprise. In practice, this means that it works on contracts with the Ministry and that ties between the two are tight. Nuklid's main office is in St Petersburg; branch offices are located in Moscow, Murmansk and Vladivostok. In 1995, the director of Nuklid was assigned the task of elaborating a programme for nuclear waste treatment, and in 1998 the organisation was appointed main contractor for the AMEC (see Chapter 2) projects and for a majority of the ten collaborative projects with Norway identified in the Framework Agreement between the two countries (see p. 3). The main Russian participant in the AMEC projects is the Russian Ministry of Defence.

On the environmental protection and control side, the most important bodies are the Federal Nuclear and Radiation Safety Authority (Gosatomnadzor) and the State Committee for Environmental Protection

(Goskomekologiya). Gosatomnadzor was established in 1991 as an executive body under the president, responsible for safety regulations in the use of atomic energy. Among its most important tasks are licensing of activities that involve the use of nuclear energy and radioactive materials, the development of standards for and monitoring of such use, non-proliferation of nuclear technology and materials, physical protection of nuclear installations, and control of Russian implementation of relevant international agreements. However, the military sector has been excluded from its brief since 1995. Goskomekologiya was the successor of the Ministry of Environmental Protection, which saw its status reduced to that of a State Committee in 1996. Its main task in the sphere of nuclear safety has been to organise environmental evaluations of various projects. In May 2000, the Russian government dissolved Goskomekologiya and transferred its main functions to the Ministry of Natural Resources.

All of the above-mentioned federal bodies have their regional representation in Northwestern Russia, where most of the projects under the Plan of Action are being implemented. The relationship between federal agencies and regional authorities is regulated through various agreements.[1] An agreement between Minatom and the regional administration of Murmansk on co-operation in the treatment of radioactive waste and spent nuclear fuel was signed in May 1998 (Murmansk Oblast 1998). A more general agreement on co-ordination of activities within the sphere of nuclear safety between Murmansk regional administration and a range of federal agencies present in the region, including Minatom, Goskomekologiya, Gosatomnadzor and the Navy through the Russian Northern Fleet, was concluded in March 2000 (Murmansk Oblast 2000). Both agreements are rather general and non-committing in nature. In 1999, a Committee for Conversion and Nuclear Radiation Safety was established at the Murmansk regional administration to co-ordinate activities in the field. This development may seem to complicate the decision-making structure further, but it is also a reminder of the significant regional implications of the various nuclear safety projects.

The most conspicuous traits in the Russian organisation of nuclear safety issues are the well-known Russian lack of horizontal integration between agencies, and the high level of conflict between them. A major line of conflict seems to run between Minatom and – in particular – Nuklid on the one hand, and the 'softer' agencies of Goskomekologiya and Gosatomnadzor, on the other. For one thing, the two latter have seen their status reduced in recent years, both formally and informally. As already mentioned, the federal environmental agency lost its ministerial status in 1996. Gosatomnadzor, for its part, has seen its major task to issue licenses threatened by new regulations that provide Minatom (in practice, Nuklid) with the right to licence activities related to the use of nuclear energy for military purposes (Government of the Russian Federation 1999). Hence, the loss of status of Goskomekologiya and Gosatomnadzor, and closure of the former, has taken place at the same time as Minatom and Nuklid have expanded their spheres of influence.

Working relations between the 'hard' and 'soft' agencies seem to be characterised by a *modus vivendi* between people who have been forced – for example, due to international projects – to maintain a certain level of contact with each other. It seems, for instance, to typify relations between the representatives of Minatom and Goskomekologiya involved in co-operation with Norway. Between others, such as Nuklid and Gosatomnadzor, a wider schism seems to prevail. There are also obvious signs of internal conflict inside Minatom. As an indication of this, it turned out at the second session of the Joint Norwegian–Russian Commission for Implementation of the Plan of Action that the head of the Russian delegation was unaware that a joint Norwegian–Russian secretariat for the Joint Commission had been set up at Minatom (Ministry of Foreign Affairs 1999b). A Nuklid employee was paid through the Norwegian project to run the secretariat. The director of Nuklid claims that not being informed about it amounted to 'simple deception'.[2] She fired the employee, and the Russian part stopped the project. The secretariat could not have been established without Minatom's knowledge and approval. Others refer to the events as 'another attempt at centralisation by [the director of Nuklid]'.[3]

A very striking feature is the negative opinions of both our Norwegian and Russian interviewees of Nuklid. On the Norwegian side, there is considerable scepticism towards Nuklid's lack of transparency in financial affairs,[4] despite acknowledgement that the organisation is able to get projects underway. But even people who try to moderate the picture use arguments such as: 'It's probably as good a "milking cow" as any other over there'.[5] Accusations from the Russian side are far more explicit. The enterprise is, among other things, accused of neglecting security standards and of harbouring intentions to monopolise all co-operation with foreigners. 'Milder' accusations refer to ineffective patterns of co-operation:

> I have nothing good to say about Nuklid. I have worked on these issues for a long time and can compare the time before and after Nuklid. Co-ordination is a good thing, but it can be done in different ways. They interfere too much with details. Before, we could co-operate bilaterally [with Norway], but now a third and superfluous structure has been introduced. They even try to take credit for projects they have done nothing but harm to.[6]

The Norwegian Plan of Action: notes from an evaluation

Co-operation at state level between Norway and Russia

Bilateral co-operation between Norway and Russia in areas covered by the Plan of Action is mainly found in the following three clusters:

- at state level between the Norwegian Ministry of Foreign Affairs and Minatom;

- through environmental co-operation, the Norwegian Ministry of the Environment and Goskomekologiya being the main participating bodies;
- in the more technical co-operation on nuclear safety between the Norwegian Radiation Protection Agency (NRPA) and Gosatomnadzor.

As to the first level, a major achievement was the bilateral Framework Agreement of 26 May 1998 (Ministry of Foreign Affairs 1998a).

The agreement states that Norway shall render free technical assistance in the stated areas, and that Russia shall exempt the delivery of such assistance from taxes, customs duties and other fees (Articles 1 and 5). Moreover, it provides important nuclear liability protections (Article 9). Ten concrete projects are identified as covered by these provisions. Among them are five of the six projects discussed below, the joint expeditions project being the only exception. The Framework Agreement foresees the establishment of a joint Norwegian–Russian commission to co-ordinate and control its implementation. The Commission has so far convened once a year and devoted most of its efforts to the implementation of the ten projects identified in the Framework Agreement. A major problem during its first years of working was the inclusion of new projects to be covered by the provisions of the Framework Agreement. This is a complex task since it has to be clarified with a range of Russian governmental agencies.

A bilateral Norwegian–Russian Commission on Environmental Affairs has been in existence since 1988. Under the auspices of this commission, an expert group on investigations of radioactive pollution in the northern areas was established in 1992 to co-ordinate bilateral activities in this field. It meets two to three times a year. A range of institutions are included from both the Norwegian and Russian sides, with Goskomekologiya, the Norwegian Ministry of the Environment and the NRPA being most heavily represented. In addition, the NRPA and Gosatomnadzor maintain continuous contact. An agreement between the two agencies on technical co-operation and exchange of information related to safe use of nuclear energy was signed in 1997 (Norwegian Radiation Protection Agency 1997). The bilateral co-operation between Norway and Russia in the sphere of environmental protection and licensing is generally characterised as very good:

> What I like in the co-operation with Norway – and I have co-operated with many countries – is that they have a broad approach instead of embracing the first and best institution they come across in Russia. They collect information before making decisions. If the decision is not always completely right, then it's at least not far from the right one.[7]

> Our co-operation with Norway contributes to strengthening our position in Russia. It's very good that Norway has taken upon itself the role of an international organiser. . . . This really helps us. [Can't you say anything critical about co-operation with Norway?] I'm sorry, but I really don't have anything critical to say. Our co-operation is indeed very

fruitful. We often come to the meetings with diverging views, but always end up agreeing.[8]

In sum, bilateral co-operation at state level between Norway and Russia in areas covered by the Plan of Action seems to have found its form although some problems and dilemmas do remain. Co-operation between environmental and nuclear safety authorities seems to function to the satisfaction of both parties. The signing of the Framework Agreement and the establishment of the Joint Commission for its implementation represent major achievements at the highest political level in the two countries. Current problems, such as those connected with the inclusion of new projects under the provisions of the Framework Agreement, are mainly to be found on the Russian side and can hardly be influenced by the Norwegian side. However, the Russians would like to see greater Norwegian understanding of the difficulties on the Russian side. To exemplify, the Framework Agreement was allegedly delayed by some six months when the Norwegian party insisted on including three additional projects after the first seven had already been approved. Also, the establishment of the secretariat for the Joint Commission at Minatom was handled fairly awkwardly by the Norwegian side (see previous section). Moreover, regional actors in Northwestern Russia complain that the Norwegian side relies too heavily on contacts with federal agencies in Moscow, especially in recent years. The same federal agencies confirm that they have nothing against direct initiatives from Norway to agencies in Northwestern Russia as long as the federal authorities are informed. Finally, there is a dilemma as to how the Norwegian side should relate to the generally unpopular organisation of Nuklid.

Project implementation

This section provides a description of how a few projects in Areas 2 and 3 of the Plan of Action were implemented. The projects were selected according to the following criteria: size in terms of financing (the total budgeted Norwegian contribution to the selected projects amounts to some 110 million NOK);[9] a certain maturity in implementation (avoiding projects that have recently been started); and institutional pluralism (attempting to cover a certain variety of institutions in Norway and Russia). In most of the projects, interviews were conduced with people at various levels in both countries. The interviews were oriented towards revealing both facts and perceptions. We explored whether goals, events and results were perceived differently in Norway and Russia. The interviews were also directed towards the actual outcome of the projects: were the established goals achieved? What particular problems were encountered? How would one characterise working relations between Norwegians and Russians in the individual projects?[10]

Effluent treatment facility for liquid radioactive waste in Murmansk

This project involved upgrading and expanding, from 1,200 square metres to 5,000 square metres, the effluent treatment facility for liquid radioactive waste at the facility serving nuclear-powered icebreakers in Murmansk. The project was conceived in 1994 as a bilateral Norwegian–Russian initiative; the USA joined the project in 1995. The facility is supposed to serve both the nuclear-powered icebreaker fleet and the Russian Northern Fleet. The opening of the facility was postponed several times but finally took place in May 2001. The Norwegian Ministry of Foreign Affairs and the NRPA were responsible for the project from the Norwegian side; main project participants in Russia were Nuklid and RTP Atomflot, on whose premises the facility was built.

A major goal of the Norwegian side was to try to convince Russia to accede to the London Convention's prohibition on dumping of nuclear waste (including low-level waste) at sea. In our interviews, this was mentioned by all Norwegians who had something to say about this project, but by none of the Russians. Instead, the latter stressed the need to solve the mounting problems of liquid radioactive waste, and to expand activities and secure revenues at RTP Atomflot (by selling services to the Northern Fleet). The Russians and Norwegians agreed that the project goals had been 'nearly achieved'. The Norwegians here probably referred to the finalisation of the facility, not to Russian accession to the prohibition on the dumping of nuclear waste in the London Convention. To explain the delays in finalising the facility, the Russian project participants referred to changes in the Russian security standards twice during the project period (in 1996 and 1998). On the Norwegian side, 'bad Russian [management] culture'[11] and 'the culture clash between Americans and Russians'[12] were referred to as factors delaying the project. Both parties mentioned the Russian need for pre-payment as a major obstacle. Nevertheless, the working relations between Norwegians and Russians on the project were generally perceived as unproblematic.

The project has been given quite a high profile and was also mentioned frequently in interviews with people not directly involved in it. Most seem to perceive it as 'partly successful'. For some time, it was clearly viewed as something close to an exemplary project since the parties had actually managed to get something done. With the repeated delays in finalising it, however, enthusiasm fell. Several of the Russian interviewees outside the project referred to it as 'too golden',[13] meaning that the abundant flows of money diminished desires on the Russian side to finalise it. Some said without mincing their words that the Norwegians should have stood more firmly on their position and demanded the finalisation of the project with the funds initially allotted.

The Lepse *project*

Lepse is the nuclear-powered icebreaker fleet's old storage vessel for radioactive waste. It is used for interim storage of spent nuclear fuel, a large proportion of which is classified as damaged fuel. This stems mainly from the nuclear-powered icebreaker *Lenin*, which suffered a reactor incident in 1966. The damaged fuel must be removed by specialised remote-controlled equipment. The vessel itself is also contaminated by radioactivity, and parts of it must be stored as radioactive waste. A Norwegian initiative resulted in the establishment of an international advisory committee, which is supposed to find a solution to the environmental threat posed by *Lepse*. The committee is headed by Norway, the other members being Russia, France, the USA, the European Commission and the Nordic Environmental Finance Corporation (NEFCO). A steering committee, composed of representatives of all donors, is also led by the Norwegian side. Responsible for the project on the Norwegian side is the Ministry of Foreign Affairs. Major participants on the Russian side include Minatom, Nuklid, Murmansk Shipping Company (the owner of *Lepse*), Goskomekologiya, Gosatomnadzor and various research institutes and enterprises.

The vessel was docked in the summer of 1999 and is assumed by Russian experts to be safe for another ten years. Documentation is being elaborated on the whole process of removing the damaged fuel and liquidating the vessel; the present project embraces only the former. There has been very little progress in the project so far, except for a fruitful sub-project on licensing. The international advisory committee and the steering committee have each met a few times. From the point of view of the Norwegian leadership of the project, practical work will not be able to start until all involved Western parties are tax exempted and assured of indemnity against liability through framework agreements with Russia. Such agreements are still pending for NEFCO and the USA. An agreement between Russia and France was signed in June 2000. The case of NEFCO is more complicated as the Russian Ministry of Foreign Affairs does not recognise NEFCO as an international organisation. The USA also needs an agreement before its participation in the project proceeds much further.[14]

Russian actors at both the project level in Murmansk and the co-ordination level in Moscow show little understanding for the Norwegian stance and complain about the lack of progress in the project:

> If nothing happens in a year's time, I'll be compelled to find other solutions. There are alternative solutions in Russia. We'll just have to go searching for money.[15]

> Why cannot Norway and Russia start the project bilaterally? *Lepse* is included in the Framework Agreement [between Norway and Russia]. Parts of the work can be started. A design has to be worked out by organisations licensed to do this, that is Russian organisations. Let's start doing that work![16]

The *Lepse* project seems to many to be the unsuccessful project *par excellence* of the Plan of Action. It is a highly profiled project involving many actors and much money, and no practical work has been done on the industrial aspect so far. The Norwegian project leadership emphasise that even with the necessary framework agreements in place a wide array of contracts will have to be drawn up between the organisations involved, and that will be no easy task either. There was a strong political wish to include *Lepse* in the Plan of Action.[17] In hindsight, it can be argued that giving such a prominent place to this complicated issue did give the Plan of Action a negative image. Some of the interviewees stressed that *Lepse* poses no radiation threat to Norway, implying that it should not have been given priority if the goal had been only to maximise Norwegian interests:

> It wouldn't have been a problem for us [in Norway] if *Lepse* had sunk in Murmansk harbour. This is a local problem, but *Lepse* became a focal case for Norway.[18]

Several Russian interviewees also expressed irritation at the Norwegian flagging of the *Lepse* project:

> *Lepse* was given top priority [by the Norwegian side], but it's become a great failure. If we hadn't docked *Lepse*, it might have sunk. I don't understand why Norway pursued this course. I don't understand how you can sit there pushing pens while nothing is being done about the ship. Do you really want to take that responsibility?[19]

In addition to providing an indication of Russian irritation at the lack of progress in the *Lepse* project, this quotation shows that the considerable Norwegian involvement in the project may have obscured perceptions of responsibility for the *Lepse* problem. Of course, the interviewee does not mean that it would have been better not to dock the vessel and let it sink. However, irritation is expressed at the fact that the Russians had to go to such measures themselves, i.e. there was a perception that the Norwegians had assumed overall responsibility for *Lepse*.

Specialised vessel for transport of spent nuclear fuel

It has long been acknowledged that Russia will need a specialised vessel for safe transport of spent nuclear fuel and possibly also other radioactive waste from decommissioned nuclear submarines from remote locations in Northwestern Russia to transfer terminals in Murmansk and the Sevmash shipyard in Severodvinsk. This is because the spent fuel is transported in containers that are too heavy to be transported by road. The ship will be required to have independent propulsion machinery, a double hull and other safety features. The firm Moss Maritime was responsible for the project on the Norwegian side; the main actors on the Russian side were Minatom and a co-ordinating body for some Russian shipyards Morskoye Korablest-royeniye (Maritime Shipbuilding).

The original plan was to build a new ship. However, in 1998 the Russian party announced that it intended instead to reconstruct an old vessel, the *Amur*, arguing that it would be a less expensive solution. The Norwegian stance was that the costs of reconstructing *Amur* would hardly be lower than building a new vessel. It also argued that *Amur* would not be able to perform the tasks of the specialised vessel. Separate expert groups were established in Norway and Russia in 1999, and, in the autumn of that year, they jointly concluded that building a new vessel would be the better alternative. At the beginning of 2000, Minatom formally informed the Norwegian party that it would go in for a new vessel. After that, planning of the building of the ship progressed rapidly.

The Norwegian project participants assumed that external concerns had delayed the project. *Amur* had constituted a problem for Minatom – how could it get rid of this old and polluted vessel? – and the planned specialised vessel for transport of spent nuclear fuel was perceived as a possible way of solving this problem. Relations between Moss Maritime and Morskoye Koroblestroyeniye were reported to be very good by both parties. The Norwegian project leader had control over all parts of the process and full knowledge of all financial dispositions on the Russian side. The Russian project representatives viewed their role as one of co-ordinating activities in Russia and to exert control on behalf of Minatom. They also considered the Norwegian approach to the project as constructive and acknowledged that the delay was due to problems on the Russian side.

Specialised railway rolling stock for transport of spent nuclear fuel

In order to transport spent nuclear fuel from terminals in Murmansk and Severodvinsk to interim storage or reprocessing in Mayak, Russia needs specialised railway rolling stock able to carry safety-approved transport containers of the type TK-18. The present project involved procurement of four such specialised wagons, of which Russia already had four. Moss Maritime was responsible on the Norwegian side; Nuklid was the main Russian contractor of the project. The wagons – finalised in March 2000 – were built at the Tver Railway Factory in Central Russia. All subcontractors were also Russian.

Negotiations on the realisation of the project commenced in September 1998. The Norwegian project manager claims that they were 'forced' by the Norwegian Ministry of Foreign Affairs to accept Nuklid as the main contractor on the Russian side. They would have preferred to control all activities in Russia, selecting subcontractors by tenders. Nuklid does not use tendering; it selects subcontractors to carry out the work at fixed prices. According to the agreement between the Norwegian project leader and Nuklid, the former had no right even to be informed of financial dispositions on the Russian side. The only reason Moss Maritime accepted it was because it was very concrete project, where progress is fairly easy to control.

The wagons were built without any particular problems. A major problem arose, however, when they were being finalised in spring 2000. The Norwegian project manager was informed that the main Russian contractor had transferred ownership of the wagons. Moss Maritime's agreement with Nuklid stated that the wagons would be owned by the Mayak facility (Moss Maritime 1998), but it turned out that they had been transferred to a newly established firm called Atomspetstrans. Moss Maritime requested further information about this enterprise and immediately stopped remaining payments to Nuklid. (The parties had agreed on twelve instalments from Moss Maritime to Nuklid.) Moss Maritime is above all concerned at the lack of respect for concluded agreements by the Russian part. They claim not to have been informed about the change of ownership, nor about the status of Atomspetstrans. Moreover, they state that, for them, it is mostly a matter of principle, but that it should be a matter of substantial worry for the Norwegian Ministry of Foreign Affairs, whose reaction to the transfer they would have preferred to be firmer. Other agencies in Russia claim that Atomspetstrans does not exist and that if it is in fact established, it will be a 'paper firm', performing only an unnecessary middle role, leasing the wagons to the Mayak facility.

A sense of considerable antagonism appears therefore to have grown between the Norwegian project management and main Russian contractor in this project:

> As far as our relationship with [Moss Maritime] is concerned, our functions are quite similar, but we're interested in the end result, and we know Russian management culture. [Moss Maritime] exaggerates its function. They're interested only in making money, which is quite understandable, but they shouldn't have been given so much money for doing this. They have only been a financial agent for the [Norwegian] Ministry of Foreign Affairs. I don't know how much [Moss Maritime] took from the Ministry of Foreign Affairs, and it really doesn't interest me, but I proposed to manage the project on my own and inform [Moss Maritime] of its progress. . . . It would have been financially beneficial also for Nuklid only to be a financial agent and receive money from [Moss Maritime] for our intellectual services, but it wouldn't have been beneficial for Russia.[20]

> [The Russians] view themselves as responsible for the project and us as donors. Nuklid thinks of itself as contractor, but they're not able to manage projects. The [Norwegian] Ministry of Foreign Affairs should support us here. They should at least insist on professional project management. The Ministry of Foreign Affairs could avoid using Nuklid by arguing that experience from other projects is not too good, for

instance in [the project on the effluent treatment facility for liquid radioactive waste in Murmansk]. . . . Norway is naive in its relation to Nuklid. If the authorities suspect that not all money is going to the agreed measures, it's wrong to continue.[21]

In sum, the project was successful in the sense that the physical product it intended to produce was in fact produced. There is, however, fundamental disagreement between the Norwegian project manager and the main Russian contractor on how such projects should be run. As a matter of principle, the former insists on their prerogative to select subcontractors in Russia, whereas the latter claims this is impossible:

These aren't commercial projects, but environmental protection. You spend far too much money on [Moss Maritime]. Now look how much cheaper things turned out in the Murmansk initiative [on the effluent treatment facility], where the NRPA was responsible. I – Nuklid – can do everything much cheaper, I'm a small firm, I'm financed by the state budget, my employees don't earn much, I can do everything much cheaper. Norwegian firms cannot evaluate the end result. If you want to reduce expenses, you should start with yourselves. You should use smaller and less expensive firms on the Norwegian side. Besides, you're not in a position to evaluate Russian firms. You need help to obtain the necessary information in a tender situation. You could be represented in the committee that evaluates the tenders, but you cannot do this alone.[22]

Upgrading of storage tanks for liquid radioactive waste at the Zvezdochka shipyard

In contrast to the preceding project, Moss Maritime was given the opportunity to carry out this one without the involvement of Nuklid. The project involved the upgrading of two tanks for liquid low-level radioactive waste of 500 cubic metres at the Zvezdochka shipyard in Severodvinsk in Arkhangelsk Oblast. The project also comprised the modernisation of piping systems and control equipment at the premises. The tank facility is located next to the site of a planned effluent treatment facility for liquid radioactive waste and will primarily function in connection with the dismantling of nuclear submarines at the shipyard. The project was started in May 1998 and completed in August 1999. As mentioned above, it was led by Norway; all subcontractors were selected through tenders in Russia. Gosatomnadzor's licensing of the tanks was developed in dialogue with NRPA.

The project is generally perceived to be one of the most successful under the Plan of Action. It involved intensive work for fourteen months and ended in the completion of the modernised tank facility. Moss Maritime used a Russian employee as project manager. He spent most of the project period in Severodvinsk. In contrast to several other projects under the Plan of Action, this one was completed without delay and at a lower cost than budgeted. In explaining the success of the project, Moss Maritime stressed its freedom to select Russian subcontractors itself, i.e. freedom from interference by Nuklid. At an early stage in the project, the Norwegian project manager discovered that payments from the Norwegian side were being used by the shipyard for other purposes than were included in the agreement. The project leader warned the Zvezdochka leadership that he would have to report this to the Norwegian Ministry of Foreign Affairs. The shipyard then decided to take up loans to pay for the approved equipment. Another problem was to make workers at the shipyard actually work. In a rather unconventional move, the project management visited the yard and promised to pay the workers a bonus if they did the work they were supposed to do. The project manager circumvented the requirement (in line with the provisions of the Framework Agreement) that inputs to the project be tax and customs exempt, simply by buying materials in the Russian market at whatever price was offered.

As mentioned, this project can also be classified as highly successful in the sense that the end product was completed by the given date. However, the director of Nuklid – who was initially appointed by the Russian side as 'leader of the project's working group' (Ministry of Foreign Affairs 1998b) – strongly opposed the objective of the project. She recognises that the project was successfully implemented, but disagreed strongly with its rationale, arguing that 'while there is no long-term gain in it for Russia, it makes it possible for commercial firms [like Moss Maritime] to earn big money in a short time – this is super-profit!'[23] She concludes that the project was 'basically unimportant, . . . far too expensive, and directed towards a completely irrelevant goal'.[24]

Analysis of material from joint expeditions to the Barents and Kara Seas

In the period 1992–1994, three joint Norwegian–Russian scientific expeditions took place in the Barents and Kara Seas. They were initiated against the background of rumours emerging towards the end of 1990 that the Soviet Union had dumped radioactive material in the Barents and Kara Seas. The rumours, which were later given credence by the Russian Yablokov report (Yablokov *et al.* 1993), led to a certain unrest in the international market for fish from the Barents Sea. Hence, it was decided at the 1992 session of the Joint Norwegian–Russian Environmental Commission to conduct joint investigations of radioactivity levels in the Barents and Kara Seas.[25] For the first expedition, Russian authorities gave permission to examine only the Barents Sea and open areas of the Kara Sea; on the next two expeditions, scientists were allowed to investigate dumping sites east of Novaya Zemlya. The expeditions revealed that the dumping had not been conducted coincidentally or recklessly, but deliberately in selected places upon the advice of radiation experts. Many objects had, for instance, been dumped in very shallow waters so that it would be possible to remove them later. More importantly, the analyses showed negligible leakage of radioactivity from the submerged material. On the basis of this conclusion, the parties agreed that the safest measure would be to leave the dumped material where it was. Only the completion of analysis work from the 1994 expedition and preparation of a collective scientific report of the three expeditions constituted a project under the Plan of Action. On the Norwegian side, the Ministry of the Environment and the NRPA bore responsibility for the project. The major participant on the Russian side was Goskomekologiya, represented by both its federal agency in Moscow and regional agencies in Murmansk and Arkhangelsk Oblasts. It should also be mentioned that representatives from the EU and the International Atomic Energy Agency (IAEA) participated in one or more of the expeditions in order to give the results international credibility.

All project participants we interviewed – at various levels and in both countries – characterised the project as extremely successful. First, the project's goals were all achieved; the expeditions were carried out and all necessary analyses performed. Second, there were no major problems in the collaboration between Norwegian and Russian actors in the project; on the contrary, the interviewees involved in the project felt that the atmosphere and co-operation between scientists and other personnel from the two countries were exceptionally good. Third, the results of the analyses have allegedly been of the utmost importance in maintaining the 'credibility' of Barents Sea fish on the international market; hence, Norway and Russia have had direct economic gains from the scientific expeditions. Fourth, Norwegian project participants emphasise that it was one of the few projects under the Plan of Action with a complex approach in terms of including an impact analysis of various possible measures.

Defining major discourses

This section places the views expressed in the evaluation of the Plan of Action in a wider empirical context. The objective is to explain why the Norwegian initiative grew to such massive proportions, how the Russian framing of nuclear safety problems affected the implementation of the initiative, and how internal criticism on the Norwegian side escalated during the time after the evaluation. Five main discourses are singled out: the 'nuclear disaster discourse' and the 'Barents euphoria discourse' on the Norwegian side, the 'nuclear complex discourse' and the 'Cold Peace discourse' on the Russian side, as well as the 'environmental blackmail discourse', which has emerged in various forms on both sides recently.

The 'nuclear disaster discourse'

The most important contributors to the Norwegian discourse on nuclear safety issues in the European Arctic have been the country's media and environmental NGOs, in particular the Bellona Foundation. Both have since the early 1990s portrayed the Kola Peninsula as a 'nuclear disaster area'. Their major achievement has been to establish an image of the nuclear installations and radioactive waste in the area as 'ticking time bombs' in the eyes of the Norwegian public and decision-makers.

The potential of the 'nuclear disaster discourse' is underscored by the general fear most people seem to harbour regarding even rumours of possible low-dose nuclear radiation.[26] As reviewed in Chapter 2, scientific data show that radiation levels in the region are low, and that the existing radiation stems from other sources than those at which the Norwegian projects are directed (e.g. Sellafield and the nuclear tests at Novaya Zemlya several decades ago). Apart from the nuclear power plant at Polyarnye Zori, there is, according to scientific information, not even any danger of nuclear contamination reaching Norwegian territory from sources in Northwestern Russia. Nevertheless, labelling the nuclear complex on the Kola Peninsula as a 'ticking time bomb' leaves people with the impression that *Lepse*, the radioactive waste and the dismantled nuclear submarines are indeed a threat to health and environment in Norway. This helps explain why the Norwegian parliamentarians were so eager to be seen to be 'doing something' about the Kola nuclear complex in the early 1990s. The complex was 'talked about' as a threat. The politicians had an opportunity to show that they could respond with determined action by assigning large sums of money to the elimination of that 'threat'. This also explains the lack of criticism of the Plan of Action from the Norwegian public. Considerable segments of the population are generally sceptical to foreign assistance programmes that have little 'in them for Norway', so the absence of criticism is surprising.[27] Clearly, the clamour created by the media and environmental NGOs surrounding the 'ticking time bomb' had more effect on people's perceptions than scientific information, leading to mounting pressures that the politicians 'do something'. To the extent that scientific information was brought to the attention of the public, people do not appear to have been convinced. 'Let's do something to clean up the radioactive waste just to be on the safe side', was the general reaction.[28]

The *Lepse* project is a good case in point. As commented by the interviewees on both the Norwegian and Russian side in the evaluation study referred to above, this project – referred to by the Norwegian researcher Edvard Stang as a 'constructed nuclear threat'[29] – was flagged as a major concern by the Norwegian government and became a symbol of the entire Plan of Action.[30] Although it is clear that the vessel constituted no threat to Norway, the *Lepse* was effective as a horrific symbol of human-made environmental degradation, urging people to think that 'something needs to be done before a nuclear disaster hits us' (see Table 4.1). The *Lepse* project is further discussed in connection with the 'environmental blackmail discourse' later.

Table 4.1 The main entities, discourse coalition, story line and metaphor of the 'nuclear disaster discourse'

Basic entities	• individuals, sub-groups in society
Discourse coalition	• Bellona, the Norwegian press, Norwegian politicians and civil servants
Major story line	• 'Something needs to be done before a nuclear disaster hits us.'
Major metaphor	• Kola Peninsula as a 'ticking time bomb'

The 'Barents euphoria discourse'

The Plan of Action, and the assistance schemes preceding it from the early 1990s, were conceived in an atmosphere of general concern in Norway about the problems experienced by the neighbouring population of Northwestern Russia, and enthusiasm to help it solve them. There was a pervading sense of optimism in this respect: the Cold War was over, the Iron Curtain gone, and the Russian northerners could – with a little help from their Nordic friends – again be included in the 'natural' brotherhood of the European Arctic peoples. The establishment of the Barents Euro-Arctic Region (BEAR) (see Chapter 2) is particularly telling: in an attempt at active region building,[31] Nordic (and, to some extent, also Russian) politicians and civil servants tried to persuade as many people as possible that the geographical area covered by the BEAR was indeed a 'natural' region. In particular, the region builders availed themselves of the metaphor of the Pomor trade that had taken place between the coastal populations of Northern Norway and Northwestern Russia (primarily today's Arkhangelsk Oblast) from around 1725 up till the Russian Revolution in 1917. A leading story line was that all northerners are 'in principle' alike, living under the same climatic conditions and sharing the same sense of distance to national centres. Furthermore, the area was now returning from the 'historical parenthesis' of Soviet isolation in the north to the 'normal' state of flourishing cross-border co-operation. The 'medicine' prescribed to foster this process was financial assistance programmes from the Nordic countries to help the Russians overcome the depressing environmental, financial and social legacy of the Cold War, and the building of new infrastructure to facilitate a broadening of contacts between East and West in the region. It was believed that the building of infrastructure would dissolve the 'cultural differences' that had developed during the seventy years of Soviet rule in Russia.[32]

The interesting thing for our discussion here is not so much whether it is 'true' that all European northerners are 'principally alike', or that new infrastructure will make 'cultural differences' disappear.[33] The point is, that until the late 1990s, it was simply politically incorrect in Norway to question these assumptions.[34] Political statements about the Barents regional development were filled with optimistic visions of reviving this geographical area as a true identity region and of the development of vigorous trade and industry links between its eastern and western parts. The media generally played from the authorities' side of the pitch, painting a picture of massive opportunities for Norwegian business and industry in Northwestern Russia. Even social science literature on the BEAR initiative was explicitly or implicitly region-building in nature.[35] The result was a 'Barents euphoria'; the leading discourse of the day established that all the problems of the European Arctic would be solved if only the Nordic countries invested in assistance programmes to help the Northwest Russian population get back on their feet.

This 'Barents euphoria' determined 'the way we speak around here' and 'the way we act around here' in Norway in the early and mid-1990s. People spoke about opportunities and played with big money. *Lepse* became a Norwegian flagship cause; defeating the contaminated vessel with damaged nuclear fuel on board would be the ultimate victory of the extended Barents regional project. In spring 2000, when 'Barents euphoria' had more or less vanished (see later on the 'environmental blackmail discourse'), the Norwegian civil servants responsible for implementing the Plan of Action tended to comment laconically on the pressures they were under to spend the money that an eager Storting had assigned: 'there was enormous pressure from the top [of the Ministry of Foreign Affairs] to get the money spent. We had to get something done while we were waiting for the Framework Agreement. It was quite horrible at times.'[36] 'I was really surprised when I first came in . . . at the lack of evaluation of the projects. . . . I don't think there's anywhere else in Norway where money has been given away that easily' (see Table 4.2).[37]

The 'nuclear complex discourse'

In spring 1990, a Norwegian television team visited Polyarnye Zori on the Kola Peninsula to investigate what the local population thought about living so close to the Kola nuclear power plant. In a clip of a classroom scene at a primary school, the teacher asks the class: 'is the power plant something sad?' 'No!' the children cry. 'Is the power plant something merry?' 'Yes!' the class retort unanimously. Ten years later, I was in Murmansk to conduct interviews for our evaluation of the Norwegian Plan of Action for nuclear safety in Northwestern Russia. Telling a Russian friend and research colleague about the mission of our present trip to Murmansk, he got slightly

Table 4.2 The main assumption, entities, discourse coalition and story line of the 'Barents euphoria discourse'

Basic assumption	• 'Natural regions are what we define them to be.'
Basic entities	• regions, sub-regions of society
Discourse coalition	• the Norwegian press and public, Norwegian politicians, civil servants and NGOs
Major story line	• 'We're all in the same boat. Let's build infrastructure!'

annoyed: 'I don't understand all this Norwegian fuss about our nuclear complex. There's nothing to worry about. Our experts know what they are doing.'[38] In one of the interviews concerning the joint Russian–Norwegian expeditions to the Barents and Kara Seas to investigate radiation levels around the dumped nuclear waste sites, our Russian conversation partner commented:

> This project was necessary for the Norwegians. Our experts knew that the ocean was clean, that the sites had been carefully selected for dumping, and that the surrounding ocean area was monitored. But that was not enough for the Norwegians.[39]

This sentiment points at crucial aspects of the Russian nuclear safety discourse: its *industrial* (rather than *environmental*) orientation (Dryzek 1997; see also Chapter 1 in this book), the perception of the *nuclear complex* as a natural entity, and the superiority of scientific knowledge to non-scientifically founded opinions. First, children in Northwestern Russia were – at least until the early 1990s – taught that the Kola nuclear power plant was 'something merry', and not 'something sad'. It was something that contributed to development and wealth in the region rather than to the degradation of health and the environment. This is, of course, a copy of the general Soviet belief in economic and social development through the conquering of the natural environment, and corresponding neglect of possible negative consequences of the industrial activities. The Kola nuclear power plant was a symbol of (Soviet) mankind's conquest of the natural world; it brought electricity, employment and economic gain to the local community, the region and the Union.[40]

The two latter aspects of the Northwest Russian discourse on nuclear safety, the notion of the 'nuclear complex' as an important and clearly defined entity on the one hand and the superiority of scientific knowledge on the other, are closely intertwined. The way in which nuclear industry is talked about as a whole reflects the vertical Soviet organisation of politics and industry. Matters of nuclear industry and safety were to be dealt with by a certain vertical structure composed of both political and industrial elements – the two were not clearly distinguishable in the Soviet Union – and these activities should not be interfered with by outside actors. The knowledge to be built on was scientific; public opinion was not a category to be taken into account, nor were the views of politicians outside the nuclear complex or of NGOs (to the extent that they existed during the Communist era). The reference above to 'our experts' – i.e. 'Our experts know what they are doing', 'Our experts knew that the ocean was clean' – is also telling. Managing a nuclear complex is a scientific task, not to be left to charlatans or the *narod* (the people)![41] An officer at the Murmansk department of the Bellona Foundation tells about his first meeting with the organisation: 'When they first came to Murmansk in 1991–1992, with offensive slogans on large banners displayed on their ship, as a Russian citizen I experienced this extremely negatively. It was such *maximisation (maksimalizm)*'.[42] The characteristics 'offensive' and 'maximisation' imply that someone has behaved improperly, interfered with something that was not his or her business.[43]

The 'nuclear complex discourse' has had a certain influence on the implementation of the Norwegian Plan of Action in some of the Russian participants' tendency to view Norwegian project partners as primarily donors, compare the co-operation difficulties between Nuklid and the Norwegian firm Moss Maritime referred to in the evaluation material above. Nuklid obviously perceived Moss Maritime as a donor straying outside its area of competence in trying to influence the implementation of the projects. Another Russian civil servant expressed similar perceptions:

> We have so many problems with the donors. They don't see the full extent of things, but jump right to individual solutions. There has to be a vertical structure. The main problem is not on the Russian side. We have settled the problems we have had in our organisation. With the donors, it's much more difficult.[44]

Labelling the Norwegian project participants as donors, it can be argued, helps the Russians to keep them at a distance. This might not have been done deliberately in order to reduce Norwegian influence on the implementation of the projects, but simply – again – because that is 'the way things are done around here'. First, nuclear safety is the responsibility of the nuclear complex; that is how it has always been and probably will remain, according to traditional Russian thinking. Second, foreign financing has come to replace the continuously decreasing federal transfers of funds to many sectors of society in the Russian regions during the 1990s, especially in gateway regions such as Northwestern Russia.[45] It is beyond the scope of this discussion to evaluate whether it is a good or a bad thing that the Russians perceive it as their natural right to carry out the projects without undue interference from 'donors'. Suffice it here to say that this opinion, rooted primarily in the Russian 'nuclear complex discourse', led to several problems in the implementation of the Norwegian Plan of Action. Likewise, the 'nuclear complex discourse' fundamentally clashes with the 'nuclear disaster discourse' that prevailed on the Norwegian side; hence, many Russians found the Norwegian preoccupation with the functioning of their nuclear complex as hysterical, scientifically unfounded and even offensive (see Table 4.3).

The 'environmental blackmail discourse'

Towards the end of the 1990s, Norwegian 'Barents euphoria' died out. Very little came out of attempts to create large Russian–Norwegian joint ventures and extended East–West trade and industry co-operation in the European

Table 4.3 The main assumption, entities, discourse coalition and story line of the
           'nuclear complex discourse'

Basic assumption	• 'Difficult questions should be dealt with by experts, not by the general public.'
Basic entities	• the components of the nuclear complex
Discourse coalition	• the Russian public, politicians and civil servants
Major story line	• 'Our experts know what they are doing.'

Arctic. One apparent success story after another ended eventually in failure.
In several famous cases, the Russians simply pressed their Western counter-
parts out when a company started to run a profit. For example, the Russian–
Norwegian wood-processing company Rossnor in Arkhangelsk Oblast was
long held up as success story number one by Norwegian authorities, but in
the mid-1990s the Norwegian partner was pressed out of the enterprise and
told in no uncertain terms to stay away from Russia. As mentioned above, it
was 'politically incorrect' in Norway to even question the declared potential
of the BEAR project in the early 1990s. Around 1997–1998, the tide turned
and the leading story line in Norwegian media was now that the entire
initiative was a great failure. 'Fiasco' became the most cited adjective to
characterise it. Compare, e.g., the following feature article in one of
Norway's leading newspapers from May 1997 (under the heading 'The
Barents Project at a Standstill'):

> Expectations were sky-high when the Barents regional co-operation
> scheme was initiated in the North Calotte after the break-up of the
> Soviet Union. A region full of optimistic inhabitants and rich in natural
> resources was to enter a new era. To get the wheels going, they would be
> lubricated by Norwegian oil money. . . . It sounds reasonable that
> Norway is spending nearly half a billion Norwegian kroner annually to
> prevent environmental disasters and foster stability in our immediate
> vicinity. Nevertheless, there is reason to ask some critical questions
> regarding Norwegian assistance. The Norwegian Ministry of Foreign
> Affairs has issued a brochure which paints a happy picture of growth and
> development resulting from our assistance programme in the east.
> Satisfaction within the Ministry does not exceed, however, that of the
> Minister of Foreign Affairs himself, Bjørn Tore Godal, who says that
> experiences are mixed. 'It's no use concealing that', says Godal. It's no
> use concealing that, roughly speaking, experience with the projects that
> involve a one-sided flow of money from Norway to Northern Russia and
> the Baltic is deemed to be good. These [projects] include the initiatives to
> stop nuclear pollution on the Kola Peninsula [*sic*!].[46] Those projects
> that require active participation by the Russians, be it in the form of
> capital, labour or simply the necessary legislation, are not moving fast, to
> put it mildly. Instances of large projects in Northwestern Russia that
> have either failed or reached a deadlock include the wood-processing

company Rossnor in Arkhangelsk, the clean-up and reconstruction of the nickel company in Pechenga, business co-operation in the oil and energy sector, and Bellona's work on nuclear safety.[47] Each of these projects entails considerable investment, both in terms of money and, not least, human energy and enthusiasm. In 1993–1994, one could go to Arkhangelsk to observe a Norwegian-run wood-processing factory that produced first-class doors and windows. This was the flagship venture of the joint Norwegian–Russian industrial effort and proof that it was possible to achieve something with a combination of Norwegian industrial competence and Russian natural resources and labour. After a few years business, the factory was simply confiscated by the Russian partner. The entire Norwegian capital of 20 million kroner is lost. Today the factory stands as a powerful warning to anyone considering investing in Russian businesses.[48]

Although it is probably the most famous case, Rossnor is not the only 'powerful warning to anyone considering investing in Russian businesses'. The Norwegian-run bakery in Nickel took over as the Ministry of Foreign Affairs's top success story demonstrating the success of the BEAR initiative, a demonstration that towards the end of the 1990s seemed to be characterised by ever diminishing enthusiasm. The bakery, too, had to be shut down around the turn of the millennium. What was once expressed by an acquaintance of mine from Murmansk, seems to have become a relevant story line in this context: 'Never believe a Russian has anything else in mind than making a fast buck when he's dealing with a Norwegian!'[49]

Co-operation in the Barents region continues. The emphasis is now on student exchange and health sector work – both clearly more successful than the industrial co-operation.[50] But no attempt has been made by Norwegian authorities to revive the 'Barents euphoria' of former days.

As mentioned in the discussion of the 'nuclear disaster discourse' earlier, the Norwegian researcher Edvard Stang was one of the earliest public critics of Norwegian assistance to the Northwest Russian nuclear sector. His starting point hinges on the assumption that Russians would be tempted to take advantage of the vast Norwegian assistance schemes:

> The *Lepse* is probably known to most people as 'that heap of rust in Murmansk'. Large quantities of high-reactive nuclear waste is stored on board, and the vessel should never have been anchored in the harbour of a large city like Murmansk. As is known, more suitable military quay structures do exist on the Kola Peninsula. But the shipowner Atomflot has discovered there is much foreign currency to be had in inviting foreign experts, television teams, well-meaning environmental activists and parliamentarians to 'inspect' this constructed nuclear threat. The director of Atomflot gladly poses for Norwegian television cameras on *Lepse*'s deck, dressed in a camel hair coat and Italian silk ties: 'Very, very

bad. If you don't help us, we will have a disaster here. Unfortunately, we have no money.' There is no reason that *Lepse* should cost Norwegian tax payers a single penny as long as Atomflot is using the population of Murmansk as 'disaster hostages' in order to profit financially. Norway has nothing to win from giving in to Russian threats of a [pending] nuclear disaster. Norwegian authorities should long ago have informed the Russians that moving *Lepse* from the Murmansk harbour is a pre-requisite for further Norwegian assistance.[51]

Robert Darst (2001) uses the term 'environmental blackmail' to describe how former Soviet republics have exploited the fact that they have on their territories installations that pose existing or potential threats to areas beyond their own borders, in order to extract as much money as possible from foreign donors. On several occasions, these states have continued polluting (or potentially polluting) activities not only because ceasing them would in itself be costly, but also because it would simultaneously imply the termination of foreign financial assistance. The case of *Lepse*, as presented by Stang, might be characterised as an example of 'environmental blackmail' although, ironically, in that case there was no real danger of a pollution disaster happening that could affect Norwegian territory; that was just something the Norwegian public in general had come to assume! In our work on the evaluation of the Norwegian Plan of Action, we often came across suspicions of environmental blackmail expressed by both our Norwegian and Russian interviewees. As already noted, a certain Russian agency in charge of the most costly construction projects under the Plan of Action was referred to by a Norwegian civil servant as 'probably as good a "milking cow" as any other over there'. Our Russian interviewees were more explicit, referring to other projects in the Plan of Action than their own. In particular, the treatment facility for liquid radioactive waste in Murmansk was generally referred to as 'too golden' for the Russian project participants to finalise it. In other words: as long as Norwegian money continued to flow across the border to meet allegedly unforeseen problems, the Russian had no incentive to finalise it. 'Surely, the project would have been finished in time had it not been for this fortunate – *blagopoluchnoye* – financing'.[52]

If the general discourse on the BEAR shifted direction – from presentations of 'success stories' to 'fiascos' – in 1997–1998, a similar change took place in the case of the nuclear safety Plan of Action around the turn of the century. Until then, the leading story line had been something to the effect that 'we help the Russians to secure their nuclear installations – look, here's a successful project!' This discourse was routinely reproduced until an 'inter-discursive transfer point' (Hajer 1995; see also Chapter 1 in this book) – moments where actors change positional statements and new discursive relationships and positionings are created – occurred in connection with our evaluation of the Plan of Action and a parallel one carried out by the Norwegian Auditor General.[53] In interviews related to these two evaluations, several Norwegian government officials said what they felt, expressing irritation at the ways in which much of the money had been spent. Shortly thereafter, the political leadership at the Ministry of Foreign Affairs joined the chorus of voices depicting the Plan of Action as a fiasco. State Secretary Espen Barth Eide provided the background information for an article in the large Oslo newspaper *Aftenposten*, headed 'Nuclear Plan Unsuccessful':

> The Ministry of Foreign Affairs is strongly critical of how Norwegian nuclear safety projects – at a price of 500 million NOK so far – have been carried out in Russia. Few of the projects have been successful. Most of the 111 projects in the Plan of Action for nuclear safety [in Northwestern Russia] have, according to the Ministry of Foreign Affairs, progressed in an unsatisfactory manner or had unsatisfactory results.[54]

Several of the projects financed by Norway were now defined as 'fiascos'. While the 'flagship venture' *Lepse* attracted surprisingly little attention as a 'failed project' – obviously, this image of an 'infected vessel' was still too scary to suggest that Norway should have avoided it – earlier 'success stories' such as the treatment facility of liquid radioactive waste in Murmansk and the specialised railway rolling stock for transport of spent nuclear fuel came under strong criticism in the Norwegian media. The waste treatment facility was referred to as a 'Norwegian nuclear crisis in Murmansk' in the newspaper *Aftenposten* in January 2001.[55] The rolling stock was characterised by the same newspaper a few weeks later as part of a 'disputed nuclear transport scheme';[56] the argument was that 25 million NOK has been spent in vain since the Mayak facility would not have the capacity to handle spent nuclear fuel from Murmansk. Both projects were also criticised by the Storting's Standing Committee on Control and Constitutional Matters in its comments on the Auditor General's Plan of Action evaluation in March 2002 (Stortinget 2002); the railway project was characterised as an outright failure.

The most famous case of alleged 'environmental blackmail' in Russian–Norwegian nuclear safety collaboration, however, relates to whether the Kola nuclear power plant should shut down according to plans or continue running as a result of Norwegian efforts to secure the power plant. The precondition on the Norwegian side had always been that assistance rendered should not prolong the life of the plant. The Russians affirmed and reaffirmed that they accepted this condition. However, Minatom decided in 2000 to prolong the life of the plant, saying in justification that the Norwegian assistance to secure the facility had made such an extension possible! The decision led to a reduction in Norwegian assistance to the power plant. The Standing Committee said further in its response to the Auditor General's evaluation (Stortinget 2002) that prolonging the life of the plant had led to an irreconcilable conflict of objectives. It was also critical of further Norwegian financing of measures at the plant before an agreement had been reached on its closure (see Table 4.4).

Table 4.4 The main assumption, entities, discourse coalition and story line of the 'environmental blackmail discourse'

Basic assumption	• 'Opportunity makes a thief.'
Basic entities	• individuals
Discourse coalition	• the Norwegian and Russian public, politicians and civil servants
Major story line	• 'Never believe a Russian has anything else in mind than making a fast buck when he's dealing with a Norwegian!'

The 'Cold Peace discourse'

At a seminar on Western nuclear safety assistance to the Russian Federation, I found myself sitting next to a top civil servant from the Russian nuclear complex at the luncheon table. Feeling slightly at a loss about what to say to him, I said something about the 'interesting talks' at the seminar. The Russian looked at me, indicating – I think – that he and I as Russian-speakers might talk less formally than he could with the other foreigners present. He said, 'You and I don't have to pretend. We both know why [Norway and the USA] are doing this. You want to destroy Russia.'[57] I think I laughed and asked him to explain what he meant. He refused, obviously convinced that any explanation was superfluous.

The wider context of the 'Cold Peace discourse' was presented in Chapter 3. This discourse does not seem to have been as prevalent in issues of nuclear safety in the European Arctic as it has in the management of marine living resources in the Barents Sea. The sentiment referred to above nevertheless shows that this type of reasoning is also to be found in matters of inter-national nuclear assistance to Russia. Although my conversation partner declined to spell out his concerns, my guess is that it was the linkage between disarmament and nuclear safety that provoked his comment about the West wanting to destroy Russia. As mentioned in Chapter 2, US nuclear safety assistance to Russia was linked to disarmament instead of environmental concerns. It is my assumption that many Russians – entangled in the 'Cold Peace discourse's' view of states as rational unitary actors (see Chapter 3) – consider it a very natural thing that US assistance is aimed more at maxi-mising US than Russian interests, compare the story line referred in Chapter 3: 'Of course, it's in Norway's interest to ruin Russia.' That one state seeks to take advantage of the 'temporary weakness' of an opponent (as expressed by Murmansk Governor Yevdokimov when commenting the *Chernigov* episode) is how things are and should be. It is also my assumption that the rationale of many of the projects under the Norwegian Plan of Action is less compre-hensible for the Russians. Why should Norway want to take care of the *Lepse* when the vessel does not actually represent a threat to Norwegian territory? Why has Norway provided such lush financing for the Murmansk treatment facility when the liquid low-level radioactive waste is also of no harm to

Norway? When the rationale of the 'opponent' – and states are, in the 'Cold Peace discourse', invariably at loggerheads – is not immediately clear, there is an arena for speculation, e.g. that the entire Norwegian nuclear safety mission is really part of a cunning and calculated plan that in the end will lead to the destruction of Russia. It could, for instance, be assumed that the closure of the Kola nuclear power plant, a prerequisite for Norwegian assistance right from the start, would constitute a 'setback' for the population of the area since alternative energy sources are not present as yet.[58] The logic of a Russian positioned in the 'Cold Peace discourse' would consequently be that 'the Norwegians are trying to fool us into going along with the closure, but we understood their poorly concealed motives, and got the better of them in the end. He who laughs last!' What could the Norwegians have done when Minatom decided to keep the power plant running? The major part of the assistance scheme had already been implemented and paid for and could no longer be withdrawn.[59]

The 'Cold Peace discourse' is also discernible in the current Russian scepticism to foreign environmental and nuclear safety engagement on Russian territory. Again, it seems as if many Russians find it hard to believe that environmental activists can have other motives than the national interests of their own countries at heart, behind a cloak of benignity (see Table 4.5). The arrest and trial of Bellona member Aleksandr Nikitin (see Chapter 1) is the best case in point.[60] Compare also the statement referred to in note 43 of Chapter 1, where I commented on the difficulties of conducting social science research in Northwestern Russia, especially in the field of the environment and nuclear safety:

> Foreign intelligence services have targeted Murmansk Oblast as a 'priority' area for their activities, Nikolai Zharkov, head of the Federal Security Service (FSB) directorate in Murmansk Oblast, told Interfax North-West on 28 December. . . . Zharkov also revealed that foreign governments frequently 'pursue their own interests' under the cover of environmental organisations.[61]

Table 4.5 The main assumptions, entities, discourse coalition and story line of the 'Cold Peace discourse'

Basic assumptions	• states are invariably in competition with each other • relations between states are zero-sum games
Basic entities	• states
Discourse coalition	• Russian politicians and civil servants; the Russian press and public
Major story line	• 'In the end, the West has only one goal: to destroy Russia.'

Discourse genealogy and power

The Norwegian initiatives to combat the potential danger of nuclear radiation in Northwestern Russia were conceived at a time of optimism concerning the possibilities of solving the environmental problems of the area. The key word of the 'Barents euphoria discourse' was *infrastructure*. It was believed to be the practical remedy necessary to clean up the European Arctic, a prime political objective of Western governments. And putting infrastructure in place required investing in numerous assistance pro-grammes. The general sense of optimism surrounding the BEAR initiative facilitated the launching of the Norwegian Plan of Action for nuclear safety in Northwestern Russia, as did the 'nuclear disaster discourse', hinging on the idea of a 'ticking time bomb' in Norway's immediate vicinity to the east. The contaminated vessel *Lepse* became a symbol of human-made nuclear pollution and the flagship of the Norwegian Plan of Action, admittedly a 'constructed nuclear threat' as far as the level of danger to Norway is concerned. The Norwegian 'nuclear disaster discourse' clashes fundament-ally with the prevalent Russian 'nuclear complex discourse' whose main assumption is that issues of nuclear safety should be left to the experts, not charlatans, environmental fanatics or the general public. Russians thinking in terms of the 'nuclear complex discourse' tend to perceive many of the Norwegian assumptions and initiatives as hysterical and offensive. They also show little understanding of the wish of Norwegians without expertise to take part in the projects, viewing them instead primarily as donors. Criticism of the Russian–Norwegian projects mounted around the turn of the century, eventually causing the Norwegian Ministry of Foreign Affairs to designate the Plan of Action as largely unsuccessful. This 'inter-discursive transfer point' took place a couple of years after a similar change had occurred in the ways the BEAR initiative was spoken of by Norwegian politicians and society at large. A leading story line regarding both initiatives was that 'the Russians are taking advantage of us', here situated within the 'environmental blackmail discourse'. There was widespread disappointment when the Russians did not keep their promise not to prolong the life of the Kola nuclear power plant. Likewise, there was mounting concern when the treatment facility for liquid radioactive waste in Murmansk was not finalised in time and cost considerably more than planned. Accusations of 'environ-mental blackmail' were already widespread on the Russian side. Participants in the bilateral projects with Norway warned that people on the Russian side were interested only in 'milking' the Norwegian public purse for as much as possible.

The 'Cold Peace discourse' is found also in matters related to nuclear safety in the European Arctic, but is not so prevalent as it was in the management of marine living resources in the Barents Sea (compare Chapter 3). Primarily, this discourse serves to obscure Norwegian motivations for many Russians: why should Norway spend large sums of money to help solve local nuclear safety problems in Northwestern Russia if the real objective is not somehow to destroy Russia? The closure of the Kola nuclear power plant, required by the Norwegian side, can also be viewed from this angle: such a closure would imply a serious blow to the local population and regional economy, a situation which would – allegedly – serve to secure Norwegian interests ('the loss of one state is always to the gain of other states').

Are there, then, any winners in the post-Cold War 'nuclear safety game' in the European Arctic? If we follow the logic of the 'Cold Peace discourse' and view states as the only relevant players on the scene, the question is reduced to a battle between Russia and Norway. Norway has obviously contributed large sums of money, the bulk of which was dispersed among various Russian actors. In return, Norway has gained a more secure Kola nuclear power plant (albeit one that will probably be in operation longer than anticipated precisely as a result of the Norwegian assistance). In addition, the first pieces of an infrastructure puzzle are in place that might enable the Northwest Russians to dispose of their radioactive waste and spent nuclear fuel, although these last two categories hardly pose any threat to Norwegian territories. In summary, Russia seems to have gained more than the Norwegians, but not as a result of deliberate discursive strategies on the Russian side; on the contrary, the prevalent Russian discourse seems rather to have downplayed the problems of the Northwest Russian nuclear complex (and hence reduced the potential for further financial assistance).[62] The winners are instead various groups on both the Russian and the Norwegian side: the Bellona Foundation succeeded in placing nuclear issues in Northwestern Russia high on the Norwegian political agenda (and, simultaneously, financing for its own activities in the region). The Norwegian media got symbolic and lucrative 'disaster stories' such as the one about *Lepse* instead of scientific reports about radiation levels. Finally, there are certain Russian organisations, notably Gosatomnadzor, that have strengthened their standing in Russian society as a result of the Norwegian assistance. The credit for the primary instance of discursive power in the case of nuclear safety seems to go to the Bellona Foundation. Availing itself of a combination of mounting 'Barents euphoria' and the potential lying in the 'ticking time bomb' metaphor – visualised by the horrific picture of the *Lepse* anchored in the harbour at Murmansk – it was instrumental in convincing Norwegian politicians and the Norwegian public that a massive assistance scheme was necessary to avoid a 'nuclear disaster'. Other discourses, such as the 'nuclear complex discourse', the 'environmental blackmail discourse' and the 'Cold Peace discourse' have clearly influenced policy outcomes, but were not to the same extent employed consciously by actors for specific aims.

5   Discourses on industrial pollution

The greatest possession of the inhabitants of Murmansk Oblast is the abundant and uniquely beautiful northern natural surroundings.
(Environmental authorities of Murmansk Oblast presenting the region on the Internet)

All the kids were healthy, except Ola – who'd been to Kola.
(Norwegian children's word game)

Introduction

As said in Chapter 2, the two nickel smelters on the Kola Peninsula – the Pechenganickel Combine at Zapolyarnyy/Nikel and the Severonickel Combine at Monchegorsk – emit large quantities of sulphur dioxide (SO_2), causing considerable acid precipitation on the Kola Peninsula and in the neighbouring Nordic countries. We also saw that emissions decreased during the 1990s as the result of a general slump in industrial output, but that considerable areas around the smelter towns have been irreversibly damaged by the pollution (see Figure 2.4). Hence, while current emission levels follow Russia's international obligations (Kotov and Nikitina 1998b; Hønneland and Jørgensen 2003), industrial pollution from the Kola Peninsula nickel smelters is still considered by Western governments to be a major environmental challenge in the European Arctic. Plans for comprehensive smelter modernisation projects had started on the Nordic side in the mid-1980s, but it was not until 2001 that agreement was reached between Norway and Russia on a project involving a 90 per cent reduction in emissions of SO_2 and heavy metals, scheduled to be finalised in 2006–2007.

This chapter is less focused than Chapters 3 and 4 on concrete management issues and project implementation; it aims more widely at defining the major Russian and Western discourses on industrial pollution in the area. To some extent, this involves assessing where in the 'problem hierarchy' of a given society matters of industrial pollution generally fit. Is the pollution discussed here – it has, by objective standards, resulted in widespread forest death – spoken about primarily as a social disaster or just as one among many problems to be taken care of by public authorities? As in the discussion about nuclear safety in Chapter 4, a main focus here will be on why the Norwegian assistance scheme acquired such massive proportions, and why it proved so difficult to implement the plans in Russia. First, it is necessary to outline the system for environmental management in Russia at the federal and regional level.

Environmental management in Russia

In Soviet times, environmental concerns – to the extent there was an awareness of them at all – were addressed in the sector ministries' five-year production plans. There was no overarching governmental body responsible for the country's environmental policy. Moreover, regional authorities, in general politically impotent, had no influence on this sector of policy. All important decisions were made along the axis running from the sector ministries at the federal level to enterprise leaderships in the regions (which, in turn, were subordinate to decisions made by the Communist Party). The main question in this section is how various federal and regional agencies have been involved in the regulation of industrial pollution of post-Soviet Russia, in particular in relation to the activities of the Kola Peninsula smelters.

The legislative framework

Soviet participation in the LRTAP regime (see Chapter 2) transferred the acid rain issue from general air-protection management to an independent element within Soviet environmental policy (Kotov and Nikitina 1998b). The first publicised Soviet national mid-term environmental programme from 1990 contained concrete requirements reflecting Soviet obligations under the LRTAP regime (Kotov and Nikitina 1998b). These have been repeated in several subsequent federal plans and programmes.

A new Soviet law on air pollution, prescribing a significant reduction in emissions and transboundary flows of the pollutants covered by the LRTAP regime, was adopted in 1982 (Organisation for Economic Co-operation and Development (OECD) 1999).[1] This law is still in effect, but after 1991 there has been a gradual revision of Soviet environmental legislation. The single most important legal act is the 1991 law on environmental protection (Russian Soviet Federative Socialist Republic 1991). The law required a shift to economic instruments in environmental policy, such as charges on polluting emissions and the establishment of environmental funds independent of the federal budget. A separate federal provision on the system for environmental funds was adopted in 1992 (Government of the Russian Federation 1992). The federal environmental fund of the Russian Federation is an independent institution responsible to the government. The regional duma of Murmansk Oblast adopted in 1997 a law on the regional environmental fund (already in existence at the time: Murmansk Oblast 1997a), and according to regional environmental authorities this fund is the most important economic environmental protection mechanism in the oblast (State Committee for Environmental Protection 2001). The income of the fund mainly comes from payments on emissions from polluting enterprises and fines for violations of environmental regulations. About 10 per cent of payments from the enterprises goes to the federal environmental fund, 30 per cent to the regional and 60 per cent to the local environmental fund of the municipality where the enterprise is located.

Regulation at the federal level

Soviet implementation of the LRTAP regime commitments was co-ordinated by a governmental interdepartmental commission, headed by the State Committee for Hydrometeorology and Environmental Monitoring (Goskomgidromet) (Kotov and Nikitina 1998b).[2] The Commission had both a territorial form of representation – including representatives of the Soviet republics that contributed to westward transboundary pollution – and sector representation from various federal agencies. It issued orders to the industrial ministries responsible for polluting enterprises. The ministries in turn set standards for emission reductions for the enterprises, incorporating them into the national plans for economic and social development. Hence, the LRTAP standards were part of a more general mechanism of political and economic government. For the most severe sources of transboundary pollution, among them the Kola Peninsula smelters, the Commission itself participated in setting the standards at the enterprise level (Kotov and Nikitina 1998b).

The federal Russian environmental agency has experienced considerable ups and downs since it was first established in 1988. As green issues came gradually to the fore in the public debate, a need arose for a political response leading to the creation of a State Committee for Environmental Protection. In 1991, the Committee was elevated to the rank of a Ministry of Environmental and Natural Resources Protection. The responsibility for co-ordinating LRTAP implementation was transferred from Goskomgidromet to the new State Committee once it was established.[3] Kotov and Nikitina (1998b) claim that the institutional framework for Soviet/Russian implementation of the LRTAP regime disintegrated as a result of this reform, arguing that the new environmental agency lacked the political authority and financial clout enjoyed by the interdepartmental Commission. This problem was compounded as the federal environmental agency gradually lost its formal status during the latter half of the 1990s, losing its ministerial status in 1996 and disbanded altogether in 2000. The agency continues as a Department for Environmental Protection under the Ministry of Natural Resources.

Regulation at the regional level

A general trait of the evolving Russian environmental legislation is that it foresees an increase in the role of regional authorities in environmental management. According to the 1993 Constitution, this area of politics is the joint responsibility of the federal and regional level. The 1997 bilateral agreement on the sharing of responsibilities between the Government of the Russian Federation and Murmansk Oblast (Murmansk Oblast 1997b: Art. 2) states that the federal level is responsible for conducting a uniform state policy in the area of environmental protection (including the elaboration and implementation of national environmental plans), while the regional level is responsible for the actual regulatory measures, for instance issuing of permits to emit polluting substances (Murmansk Oblast 1997b: Art. 4).

The Murmansk regional administration does not have a department for environmental regulation. Rather, the regional representation of the State Committee/Department for Environmental Protection (in the following referred to as the regional environmental committee) functions as an imple-menting agency not only for its superior federal office in Moscow, but also for the regional administration. The regional administration determines the oblast's environmental policies by elaborating programmes, action plans and concrete regulations. The annual report of the Murmansk regional environ-mental committee (State Committee for Environmental Protection 2001) states that regional environmental legislation is largely the business of the regional administration and only partly of the regional duma. In the case of air pollution control and environmental management more generally, there has indeed been a certain transfer of responsibility to the regional level. However, the regional administration has not found it necessary to establish a department for environmental protection within its structure; it has con-tinued instead to work with the regional environmental committee, which, formally speaking, is a representation of federal authorities located in the region.

In addition to the elaboration of environmental programmes and regu-latory standards, as well as the monitoring and enforcement activities performed by the regional environmental committee, the regional authorities are responsible for the regional environmental fund. The money from enterprises as polluting fees or as fines is first deposited into a special account controlled by the regional environmental committee. From there, 10 per cent is distributed to the federal environmental fund, 30 per cent to the regional environmental fund and 60 per cent to the local environmental funds of the oblast (Murmansk Oblast 1997a). The municipalities where the Kola Peninsula smelters are located, Monchegorsk and Pechenga rayons, received 30.1 and 34.7 per cent respectively of this money in 2000 (State Committee for Environmental Protection 2001). Further, the regional fund in 2000 contributed financially to five environmental programmes at the regional level. One of them was aimed at the restoration of the environment around the Pechenganickel and Severonickel smelters. The programme was awarded

329,000 roubles from the fund in the period 1997–2000 (State Committee for Environmental Protection 2001).

Defining major discourses

The environmental degradation around the Kola nickel smelters is a material fact. The organisation of federal and regional regulation systems, as well as international regimes, to combat further damage to the environment is an institutional fact. Our main objective in the following is to discuss how the environmental consequences of the smelters are talked about in Russia and the Nordic countries, and how this affects institutional arrangements at international, federal and regional levels. Two new discourses are defined: the 'death clouds discourse' in Norway and the 'anti-hysteria discourse' in Russia. In addition, two of our old acquaintances from Chapters 3 and 4 are encountered in relation to issues of industrial pollution: the 'Barents euphoria discourse' combined with the 'sustainability discourse' on the Nordic side, and the 'environmental blackmail discourse' found on both sides of the border.

The 'death clouds discourse'

Passing Monchegorsk by car on our way from Apatity to Murmansk recently, a Norwegian travelling with me in the car said, 'When I came to this area for the first time in the late 1980s, it was simply impossible to absorb the extent of environmental degradation. I could only imagine that this was what a nuclear war zone would look like'.[4] In 1990, the Norwegian environmental NGO Nature and Youth made the following statement: '[Kirkenes] will have become a desert in 20 years!'[5] In an article on the Russian nickel industry, Andrew Bond (1996) writes:

> Anecdotal accounts are sufficient as an indication of the general magnitude of the problem. . . . On the Kola Peninsula, forests within a 20-km radius of the city of Monchegorsk were reported to be completely dead, with vegetation stress detectable as far as 60 km away in Apatity and across international borders in Finland . . .; at Nikel', nearly half the workforce was reported to be suffering from respiratory ailments . . ., life expectancy is estimated at 42 years, and a 'black desert' spreading outward from the Pechenganickel smelter was wreaking sufficient havoc across international borders to compel Norwegian officials to undertake initiatives to reduce emissions at the Combine.
>
> (Bond 1996: 307)

Words used to describe the European Arctic environment in these extracts are 'nuclear war zone' and 'black desert'. The most famous characteristic is, however, the metaphor of the 'death clouds' drifting across the border from the Kola Peninsula to the Nordic countries. In the late 1980s, the citizen action campaign that called itself 'Stop the Death Clouds from the Soviet Union' was founded in Kirkenes, the Norwegian town closest to the Soviet/Russian border.

> 'No-one wants to live in a desert', says [leader of the campaign]. On behalf of the campaign, he demands the necessary 200 million kroner to be taken from the defence budget. He says that the Cold War has now been replaced by a Hot War, the war against the environmental threat. . . . 'This money should be spent on environmental defence', he demands on behalf of the campaign, which includes people as diverse as directors, housewives, teachers and local politicians. Previously, the actionists crossed the Soviet border and chained themselves together [at the Norwegian–Finnish–Soviet border] in protest about the emissions from Nikel.[6]

From the end of the 1980s, pictures of the blackened tree stumps in the areas surrounding Nikel and Monchegorsk became perhaps the most widely used 'horror shots' in the Norwegian news media. The Kola Peninsula was depicted as a 'moon landscape', 'black desert' and so on, in short, a place unfit for human habitation. When, in the mid-1990s, I was lecturing on international Arctic relations at the University of Tromsø, I had fun each new term telling students about the wonderful fishing rivers and ski resorts of the Kola Peninsula (see next section for this alternative to the 'death clouds discourse'). A large proportion of the students always found it hard to believe me; obviously, it was difficult for them to picture the Kola Peninsula as anything else than a wasteland of nuclear radiation and billowing 'death clouds'. The absurd and sometimes macabre 'all-the-kids jokes' – popular among Norwegian children for decades – had their own Kola variant: 'All the kids were healthy, except Ola – who'd been to Kola'.[7] Or, as expressed rather emotionally in a review of a Norwegian book on everyday life in Nikel (the book itself entitled 'Under the Three Pipes'),

> How many times have I not heard about the 'death clouds' from Nikel, but who has told or written that *within* these death clouds masses of Russians have lived and been forced to work to keep it going from day to day? . . . I ask myself: how can they protect themselves? How can they live a decent life?[8]

The environmental degradation around the Kola nickel smelters is indisputable; see Figure 2.4 for a visualisation of its extent and gravity. The effects of the sulphur emissions on human health are more disputed; Bond's (1996) 'anecdotal accounts' referred to above are clearly exaggerated. Several joint Russian–Norwegian health studies of the inhabitants of Northern Norway and Northwestern Russia conclude that the effect of the pollution is not significant.[9] The depiction of the Kola Peninsula as one big 'moon landscape' or 'black desert' is positively incorrect. The black tree stumps are in the immediate vicinity of Nikel and Monchegorsk; the larger parts of the peninsula are clean and practically untouched wilderness, with landscapes varying from river deltas to ranges of alpine mountains. Nevertheless, these tree stumps are what are 'talked about' in the Norwegian media and by the public in general – to such an extent that most Norwegians seem to find it difficult to realise that it is not 'the whole picture' as far as the environment of the Kola Peninsula is concerned. It was also the tree stumps that spurred the Norwegian government to set aside in 1990 300 million NOK to modernise the Pechenganickel smelter (see Table 5.1).

The 'anti-hysteria discourse'

If you visit the home page of the regional environmental committee of Murmansk Oblast on the Internet, you will be welcomed by the following words, rendered like a Soviet-type slogan in the upper-right corner of the page: 'the greatest possession of the inhabitants of Murmansk Oblast is the abundant and uniquely beautiful natural surroundings'.[10] The first page mentions the 'strong anthropogenic (*antropogennyy*) influence' on the region's environment, but devotes more space to elaborate on the 'abundance of untouched corners of nature' (including the 'thousands of flocks of eider-ducks' to be observed in the Kandalaksha Bay and the 'many thousands of pregnant Greenland seals' on the ice at the mouth of the White Sea in springtime). I have repeatedly encountered this perception of the inhabitants of Murmansk Oblast of their immediate physical environment: the Kola Peninsula as a pristine corner of the world, blessed with many natural assets and the purest water and air. Confronted about pollution from the nickel smelters, these Russians would often answer to the effect that, 'Yes, there are polluted spots in our region, but most of it is utterly uncontaminated'. And, by objective measures, they are right: some areas of the Kola Peninsula are heavily polluted, but the major part is not. Where the Nordic 'death clouds discourse' (and, for that matter, the 'nuclear disaster discourse' presented in Chapter 4) uses the black tree stumps as symbolic of a Kola Peninsula as a wasteland of nuclear radiation and dead forests, the Russians focus instead on the environmentally pure character of the larger parts of their region.

Table 5.1 The main entities, discourse coalition, story line and metaphors of the 'death clouds discourse'

Basic entities	• individuals
Discourse coalition	• the Norwegian press and public, politicians, civil servants and NGOs
Major story line	• 'In 20 years, Kirkenes will have become a desert!'
Major metaphors	• the Kola Peninsula as a 'moon landscape', 'nuclear war zone' and 'black desert'

In Chapter 3, I mentioned the complaints of the leader of the Barents Press office in Murmansk that Nordic journalists are interested only in 'disaster stories' from Northwestern Russia. The interview with her was made for an evaluation of media and competence projects carried out under the auspices of the BEAR agreement (Jørgensen and Hønneland 2002). Among these projects were several courses and seminars held in Norway for Russian journalists, aimed at teaching them to work in a 'free press' climate. The project participants we interviewed for the evaluation were generally satisfied with the courses and seminars, but they objected to the thinly veiled attempts of Norwegian course leaders to teach the Russians to do things 'the Norwegian way'. It would be impossible anyway due to the 'different mentalities' of the two countries, according to our interviewees. One of the things that tired them was the Norwegians' urging to criticise companies on the Kola Peninsula that polluted the environment. As one of the journalists put it: 'Russia has big [environmental] problems. We could have criticised and berated [the polluters], but what good would it have done?'[11] His point was that the mono-industrial towns on the Kola Peninsula are totally dependent on the local industry. Everybody knows that the enterprises are polluting, but without them there would be no towns at all. 'Yes,' I can imagine my interviewee proceeding, 'we do have a pollution problem, but what good does it do going all hysterical?'[12]

Another interesting feature of the Northwest Russians' perception of their environment is where they rank industrial pollution in their 'hierarchy of problems'. It is a well-known fact that environmental concerns vanished from most people's agenda during the 1990s as the economic crisis of the country affected their lives to an increasing extent.[13] In addition, there are some peculiarities related to the Russian perception of 'life in the north' that, for many, seem to upstage the problems of industrial pollution in the region (besides their pretty unruffled approach to pollution in the first place). It is particularly interesting to observe the difference between Russian reactions to the post-Soviet economic and social crisis in Northwestern Russia and Norwegian reactions to earlier economic recessions – admittedly less severe – in Northern Norway. While the more or less explicit slogan on the Norwegian side of the border is 'this is where we belong!' the corresponding device on the Russian side seems to be 'this is where we have sacrificed ourselves for the Motherland!' Where Northern Norwegians in times of crisis send the following message to the national centre: 'Help us to stay!', Northwest Russians – at least on the Kola Peninsula – cry out: 'Help us out of here!' The numerous programmes for 'depopulation of the North' currently in operation in Northwestern Russia are telling in this respect.[14] Furthermore, 'life on the Kola Peninsula' is perceived very differently in the Nordic and Russian parts of the European Arctic. In the Nordic countries, as already mentioned, the general view of the Kola Peninsula is above all one of a seriously polluted area; one feels sorry for those who are forced to live under such environmental conditions (see the previous section on the 'death clouds discourse'). On the Russian side of the border, however, the main concern

Table 5.2 The main discourse coalition, story line and metaphor of the 'anti-hysteria discourse'

Discourse coalition	• the Russian press and public, politicians and civil servants
Major story line	• 'Yes, we do have a pollution problem, but what good does it do going all hysterical?'
Major metaphor	• the Kola Peninsula as a pristine corner of the world, blessed with bountiful natural riches and the purest water and air

seems to be the climate: people feel sorry for themselves because they have to live under such climatic conditions! There is a general view of life in the Arctic as a great sacrifice, and hence praiseworthy, but by no means 'natural'. For instance, while participating in a study of public child care on the Kola Peninsula (Berteig *et al.* 1998), I often heard officials in the region's child-care system complain that it is not natural for children to grow up under Arctic climatic conditions. A main concern was the lack of money to send children in care out of the region more often.[15] Many child-care institution officials mentioned the 'unnatural' climatic conditions as a main cause of illness among children.[16] Fear of industrial pollution – e.g. its effects on the children's health – was not mentioned once (see Table 5.2).

The 'Barents euphoria/sustainability discourse'

The 'Barents euphoria discourse' was elaborated in Chapter 4. It is rooted in the region building efforts – 'talking a region into existence' – put in place by Nordic, and particularly Norwegian, authorities since 1993. Also, it hinges on Nordic people's willingness to help their 'poor' neighbours in the east, a trait that emerged in the late 1980s. The building of infrastructure was believed to foster more business collaboration and exchange of people between East and West in the region, which in turn was supposed to break down the 'cultural differences' that existed as a result of seventy years or so of Soviet rule in Russia.

In questions of environmental management, the 'Barents euphoria discourse' – reflecting the rather uncritical belief that money, infrastructure and exchange of people could solve the problems of Northwestern Russia – is closely linked to the 'sustainability discourse', presented briefly in Chapter 1 and more elaborately in its 'fishery variant' in Chapter 3. A striking feature of the entire BEAR initiative is that questions about 'the environment' are supposed to permeate the entire collaborative venture, spanning from issues such as culture and tourism to indigenous peoples and science. As expressed in the Kirkenes Declaration, initiating the BEAR co-operation:

> The Participants emphasized that the environmental dimension must be fully integrated into all activities in the Region, inter alia, through the establishment by the states in the Region of common ecological criteria

for the exploitation of natural resources and the prevention of pollution at source and recognized that solving the existing major transboundary environmental problems will be important in realizing the potential for broader co-operation in the Region.

(Ministry of Foreign Affairs 1993: 2)

Hence, environmental concerns – i.e. the principle of sustainable development – are incorporated into all aspects of the BEAR programme, from industrial projects to culture and the rest of the ten prioritised areas.[17] In joint work on cultural matters for the period 1995–1997, the single largest project (allotted 5 million NOK out of the 13.7 million NOK made available) was devoted to environmental issues, 'Environmental Challenges in the Barents Region – also a Question about Culture'. The project is described in a background document formulated by the BEAR working group on culture:

> Background:
> Solving environmental problems is one of the main objectives of the Barents co-operation. It is advisable, therefore, that one of the [cultural] projects should have as its general focus this environmental challenge. By means of artistic and cultural activities, people in the region could take on a new and motivating approach to the concept of the environment. Not least, the environment should be emphasised as a *positive* factor in people's life in the Arctic.
>
> Goals:
> - to increase awareness and knowledge about environmental problems in the region
> - to enhance environmental involvement of the people of the region
> - to mobilise the region's cultural and creative forces in a concerted effort to focus on the environmental challenges of the region
> - to show how the environment can be an enriching factor for people living in the Arctic.
>
> (Hønneland 1994)[18]

The project involved a variety of cultural activities, from exhibitions to puppet shows about people and the environment. In hindsight, the combining of environmental concerns with cultural activities seems rather contrived and slightly comical. It was, however, 'the way one spoke around here' in the early and mid-1990s. Nearly a decade later, the 'Barents euphoria' itself having more or less died out, the forced linking of environmental issues to all other aspects of the BEAR initiative had also disappeared. The amount of money channelled into the 'Question about Culture' project is amazing from the perspective of the 2000s: the 5 million NOK spent on a project involving mainly environmental education through song and dance represents a quarter of recent years' annual budgets to the entire BEAR programme (see Table 5.3).

Table 5.3 The main assumption, entities, discourse coalition and story line of the 'Barents euphoria discourse'

Basic assumption	• 'Natural regions are what we define them to be.'
Basic entities	• regions, sub-regions of society
Discourse coalition	• the Norwegian press and public, Norwegian politicians, civil servants and NGOs
Major story line	• 'Co-operation in the Barents region – always a question about the environment!'

The 'environmental blackmail discourse'

The arguments of the 'environmental blackmail discourse' on nuclear safety, presented in Chapter 4, are also recognisable in discussions about industrial pollution in the European Arctic, and in particular in relation to Norwegian support for the modernisation of the Pechanganickel smelter. Since the modernisation agreement was signed in June 2001, a recurring theme in the Norwegian press has been suspicion that the allotted 270 million NOK either will not be used for the designated purpose, or will be used to prolong the life of the smelter, as expressed for instance in an editorial in a local newspaper in Finnmark, the Norwegian county bordering on Murmansk Oblast:

> We think it would have been more sensible to use the money on other environmental measures in the western parts of the Kola Peninsula. Prime Minister Jens Stoltenberg has been to Russia with a check of 270 million NOK. The money is meant to clean up the emissions from the nickel smelter. Several people are critical of the fact that the Norwegian government is giving these millions to a company that last year ran with a surplus of 28 billion NOK. The management of the Norilsk Nickel Combinate could pay a dividend of nearly 17 billion NOK to its limousine-driving stock holders. . . . What will now happen, is that the smelter's life will be prolonged indefinitely, the owners are not willing to do anything about the pollution – even if the Norwegian million kroner gift is placed in an account in a Russian bank. Because that money can easily be spent on other initiatives, initiatives that have nothing to do with the environment.[19]

The phrase 'limousine-driving stock holders' of Norilsk Nickel resonates with the description in Chapter 4 of the director of Atomflot, 'dressed in a camel hair coat and Italian silk ties'. It reflects the suspicion that Norwegian assistance is not really necessary, and that there is a high risk that it will be diverted and spent on personal consumption by already wealthy Russians instead of environmental measures. Such suspicion is even more prevalent on the Russian side, compare the story line referred to the 'environmental blackmail discourse' in Chapter 4, 'Never believe a Russian has anything else in mind than making a fast buck when he's dealing with a Norwegian!' This discourse also reflects the scepticism felt about the development of the Russian system for environmental management, focusing on the fact that enterprise directors in Russia often have little incentive to make environmentally sound investments due to the insecurities inherent in Russian economic life in general and the Russian property system in particular. Further, the fall in production since the break-up of the Soviet Union renders modernisation projects unnecessary; the Russian Federation is already in compliance with its international obligations regarding industrial pollution (see Table 5.4). As Vladimir Kotov and Elena Nikitina, two of Russia's most well-reputed researchers on issues of industrial pollution, say:

> At the most basic level, the Norilsk case indicates that Western environmental policy instruments will not necessarily be effective in a country undergoing a major economic and political transition. Corruption, lack of motivation, and the inability to impose sanctions on those who violate the law all work against attempts to solve environmental problems through regulation. Compounding the problem, of course, is the absence of effective ownership in Russian industry. . . . It is entirely feasible technologically to refurbish Norilsk's plants on the Kola peninsula so that they will be less polluting. The most important impediment to this is the unsettled state of property rights in Russia, which eliminates the incentive to make environmental investments. Remove that impediment and the way will be open for such investments, both by individual firms and through joint implementation projects. Of course, there are those who oppose such an approach, who would prefer simply to close highly polluting facilities. . . . [T]here is no compelling reason for the Russian government to shut Norilsk's Kola facilities down: The country is technically meeting its international obligations under LRTAP with respect to sulfur dioxide emissions, and closing the facilities would entail high economic and social costs. . . . A significant negative development is the elimination of Russia's Ministry of the Environment. From now on, environmental protection in Russia will be in the hands of a state committee rather than a full ministry.[20] This means that far less emphasis will be given to the environment because the chair of the committee will be excluded from high-level decisions affecting it. How these developments will ultimately play out is difficult to say, but those wishing to protect Russia's environment would do well to recall the survival strategy of the gulag prisoners who originally built Norilsk Nickel: 'Trust nobody, fear nothing, never beg.'
>
> (Kotov and Nikitina 1996: 35–37)

Discourse genealogy and power

The industrial pollution from the Kola Peninsula nickel smelters is above all an issue at the interface between the Nordic 'death clouds discourse' and the

Table 5.4 The main assumption, entities, discourse coalition and story line of the 'environmental blackmail discourse'

Basic assumption	• 'Opportunity makes a thief.'
Basic entities	• individuals
Discourse coalition	• the Norwegian and Russian public, politicians and civil servants
Major story line	• 'Never believe a Russian has anything else in mind than making a fast buck when he's dealing with a Norwegian!'

Russian 'anti-hysteria discourse'. The Kola Peninsula has been designated by the Nordic public as a catastrophe area, a 'moon landscape', a 'nuclear war zone or 'black desert' – even small children 'know' that you get sick if you visit Kola! Although Russians generally acknowledge that the nickel smelters cause considerable pollution, they tend to rank it lower in their 'hierarchy of problems' than their Nordic neighbours. First, they emphasise that the Kola Peninsula is for the most part a pristine area of the world, blessed with many natural assets and the purest water and air. Second, they tend not to 'go all hysterical' about pollution that does exist. They admit that it is a bad thing, but point out that there are many other bad things in life and, in addition, new problems would be created if the smelters were forced to close down. The 'Barents euphoria discourse' of the early and mid-1990s – combined with the 'sustainability discourse' permeating the entire BEAR venture – resulted in something of a song and dance about 'our common future' and so on. The euphoria has faded, and while the dancers and singers still dance and sing, it is no longer about the environment. The Russians have finally managed to devise a cleaning scheme for the Pechenganickel smelter that has convinced the Norwegians to pay out the larger part of the money set aside for the modernisation project as early as in 1990. However, the 'environmental blackmail discourse' had already made a place for itself in the mind of the Norwegian public, and the original enthusiasm for this grand environmental measure is now largely replaced by a nagging idea that the money will go to the 'limousine-driving stock holders' of Norilsk Nickel, and that as well running a company as Norilsk Nickel should be made to pay for its own modernisation.

The exercise of power in this case lies above all in the pursuit of the 'death clouds discourse' on the Nordic side, followed most vociferously by Nordic environmental NGOs and their supporters in the political arena. Well assisted by the media, they succeeded in creating an image of Norway's eastern neighbour in the north as a catastrophe zone that had to be saved by introducing new infrastructure, i.e. money. Since the money had already been set aside in the 1990 Norwegian state budget, the ensuing 'environmental blackmail discourse' – not to speak of the 'anti-hysteria discourse' on the Russian side – never attained the momentum to prevent the realisation of the modernisation project once the Russians said 'jump'.

6 Conclusion

Introduction

I started out in Chapter 1 by saying that this book is about how we talk about the environment, why we talk about the environment in a certain way, and some of the effects of doing so. It has not been my ambition to explain all aspects of the management of marine living resources, nuclear safety and industrial pollution in the European Arctic. It *has* been my ambition, however, to explore how discourses 'define the range of policy options' (Litfin 1994) in these issues. How are the borders defined in the three case studies of what is to be perceived as legitimate knowledge, actor (including national) interests and institutional arrangements? In this discussion, I have drawn partly on Litfin's (1994) conception of how knowledge is brokered by different people and groups, partly on Hajer's (1995) ideas of story lines and the positioning of subjects, and, to some extent, also on Dryzek's (1997) categorisation of various environmental discourses. My major focus has been on the embeddedness of environmental discourses in more overarching discourses in society (Neumann 2001).

The three case studies are not directly comparable, for several reasons. First, their subject matter varies. Where the study on marine living resources deals primarily with the division of natural resources between Norway and Russia – with the potential for considerable economic gain for each of the parties – the issues of nuclear safety and industrial pollution are basically questions regarding who should shoulder the burden of remedying environmental damage, or preventing future damage. In turn, the two latter cases vary in the degree to which they have so far caused harm beyond the borders of the Russian Federation, i.e. on the territories of the Nordic countries.

Second, the empirical data employed in the three case studies are not directly comparable, either. All three build on my general knowledge of environmental affairs in the region through (partly participant) observation (see the section on methodology in Chapter 1). In addition, the study on marine living resources draws heavily on the abundant material found in both Russian and Norwegian media about the management of the Barents Sea fisheries since the late 1990s. The chapter on nuclear safety is based partly on my participation in an evaluation of the Norwegian Plan of Action on nuclear safety in Northwestern Russia, which gave direct (interview) access to key players on both the Russian and Norwegian side. The empirical data on industrial pollution is, conversely, less abundant since it has never been as hot an issue as nuclear safety and, in particular, fisheries management in recent years.[1] The time aspect is also relevant since the study on marine living resources, for instance, would have given quite different results had the investigation been performed a few years earlier, when cod was abundant in the Barents Sea and the allocation game between the parties hence easier to solve. In conclusion, the objective has been not so much to compare the three case studies, but to let each of them contribute more generally to elucidating the environmental interface between Russia and the West, Norway in particular, in the European Arctic.

This concluding chapter starts out with brief summaries of each of the three case studies, focusing on the genealogy of each cluster of discourses. It continues to discuss how knowledge brokerage and the employment of story lines and metaphors affect discourses, and how environmental discourses are embedded in more general discourses. The discussion is rounded off with further remarks of a general nature concerning how clusters of discourses in Russia and the West have affected environmental issues in the European Arctic since the early 1990s.

Discourses on marine living resources

The management of the Barents Sea fish stocks is a bilateral responsibility shared by Russia and Norway. Since the mid-1970s, the Joint Fisheries Commission has established TACs for the three joint stocks in the area: cod, haddock and capelin. Of these, cod is by far the commercially most important. Towards the end of the 1990s, the cod stock was declining at an alarming pace. At the same time, the internationally acknowledged precautionary principle – which says that lack of scientific certainty should not be used as a reason not to undertake management measures that could prevent degradation – was formally accepted by the scientific community in ICES and the Joint Fisheries Commission. The marine scientists recommended drastic reductions in the Barents Sea cod quota, but the Joint Fisheries Commission annually established quotas far above these recommendations. The Russian party to the Commission strongly opposed the need for implementing quota reductions. The Norwegian party generally supported the scientific recommendations although opinions varied within the Norwegian fishing industry.

We singled out four major discourses pertaining to the Barents Sea fisheries management: the 'sustainability discourse' and the 'pity-the-Russians discourse' on the Norwegian side, the 'Cold Peace discourse' in Russia, and the 'seafaring community discourse', which is found on both sides. The Norwegian discourse on fisheries management in the Barents Sea has been a very vocal rehearsal of the 'sustainability chorus'. Whatever the issue, i.e. disregard for scientific opinion or stricter Norwegian enforcement in the Svalbard Zone, excuses such as 'it's still within sustainable limits' have calmed ruffled feathers and explanations have evoked 'the need for sustainability'. When the Norwegians started arguing in 1999 in the Joint Fisheries Commission that quotas needed to be cut, and that the concern was one of sustainability, Russian suspicions were aroused. The Russians generally think that cod stock indications are not really too bad, and that ICES reference points are consequently too high. The establishment of cod TACs is certainly *not* a matter of sustainability in their eyes, which is why they suspect a hidden agenda in the Norwegian stance. The dominant 'Cold Peace discourse' in Russia provides an arena for speculation. Seen through the lens of this discourse, states are ceaselessly at loggerheads over material resources, and it is always in one state's interest to damage the interests of another (even when no immediate gains for oneself are apparent). In addition, the Russians have 'observed' that the West has been consciously trying throughout the 1990s to ruin Russia under the cover of 'democratis-ation' and 'economic restructuring'. The Norwegian 'sustainability chorus' is easily recognisable as a foil defending Western national interests, either in the form of maintaining high world market prices for cod or simply damaging Russia as a competitor. Story lines such as 'every country defends its own interests with the means available to it' and 'Norway does everything it can to destroy the Russian fishing industry' give sense to otherwise unclear motivations.

At the same time, the 'seafaring community discourse' feeds on distrust on both sides of the border of the scientists' pessimistic prognoses for the Barents Sea cod stock. The discourse portrays the science as fundamentally out of step with the experiences of the people at sea, fishers and other types of sea folk. Scientific language is viewed with disdain, and the 'seafaring com-munity discourse' seizes any opportunity to put out claims that the science is wrong. Its main objective is to give credence to the notion that fishing should be left to the fishers; experts are not needed. Its most important consequence is to weaken the scientific view that the Barents Sea cod stock is in danger of overfishing by inflating the margins of uncertainty that always accompany scientific prognoses. Its purpose, in other words, is to weaken 'sustainability discourse' arguments and strengthen the conclusions (if not the arguments) in the 'Cold Peace discourse'.

Divergent views on how to manage the fishery seem to have reached deadlock: the Norwegian 'sustainability discourse' and the Russian 'Cold Peace discourse' are based on incompatible premises. The 'seafaring community discourse' supports Russian argumentation, reducing, as it does so, the chances of the Norwegians to win support. The 'pity-the-Russians discourse' offers a way out of the deadlock: the Norwegians are ready to give in on (unreal but probably sincerely felt) humanitarians grounds; the 'difficult circumstances of the Northwest Russian people' means that all good sustainability intentions must be dropped. The result is a type of management which, by definition, flouts the internationally recognised precautionary approach.

Discourses on nuclear safety

Radioactive pollution in Russia and Eastern and Central Europe has been designated one of the major environmental and security policy challenges in the European Arctic region in the post-Cold War period. There is widespread nuclear activity in the area, both civilian and military, particularly in Northwestern Russia. Hazards stem from unsatisfactory storage of large quantities of radioactive waste, decommissioned nuclear submarines awaiting dismantling, and the continued operation of unsafe nuclear power plants. Extensive Western efforts, in particular from Norway and the USA, have aimed at reducing the threat of the potential spread of radioactive pollution from Northwestern Russia since the early 1990s.

In this case study, five major discourses were defined: the 'nuclear disaster discourse' and the 'Barents euphoria discourse' on the Norwegian side, the 'nuclear complex discourse' and the 'Cold Peace discourse' on the Russian side, and the 'environmental blackmail discourse', which has emerged in various forms on both sides. The Norwegian initiatives to combat the potential danger of nuclear radiation in Northwestern Russia were conceived at a time of optimism concerning the possibilities of solving the environmental problems of the area. The key word in the 'Barents euphoria discourse' was *infrastructure*. It was believed to be the practical remedy necessary to clean up the European Arctic, a prime political objective of Western governments. And putting infrastructure in place required investing in numerous assistance programmes. The general sense of optimism surrounding the BEAR initiative facilitated the launching of the Norwegian Plan of Action for nuclear safety in Northwestern Russia, as did the 'nuclear disaster discourse', hinging on the idea of a 'ticking time bomb' in Norway's immediate vicinity to the east. The contaminated vessel *Lepse* became a symbol of human-made nuclear pollution and the flagship project of the Norwegian Plan of Action, admittedly a 'constructed nuclear threat' as far as the danger to Norway was concerned. The Norwegian 'nuclear disaster discourse' clashes fundamentally with the prevalent Russian 'nuclear complex discourse' whose main assumption is that issues of nuclear safety should be left to the experts, not charlatans, environmental fanatics or the general public. Russians thinking in terms of the 'nuclear complex discourse' tend to perceive many of the Norwegian assumptions and initiatives as hysterical and offensive. They also show little understanding of the wish of ordinary Norwegians to take part in the projects, viewing them instead primarily as donors. Criticism of the Russian–Norwegian projects mounted around the turn of the century, eventually causing the Norwegian Ministry of Foreign Affairs to admit that the Plan of Action had been largely unsuccessful. This 'inter-discursive transfer point' took place a couple of years after a similar change had occurred in the ways the BEAR initiative was spoken of among politicians in Norway and society at large. A leading story line regarding both initiatives was that 'the Russians are taking advantage of us', a statement typical of the 'environmental blackmail discourse'. There was widespread disappointment at the Russians' failure to keep their promise not to prolong the life of the Kola nuclear power plant. Likewise, there was mounting concern when the treatment facility for liquid radioactive waste in Murmansk was not finalised in time and cost considerably more than planned. Accusations of 'environmental blackmail' were already widespread on the Russian side. Participants in the bilateral projects with Norway warned that other people on the Russian side were interested only in 'milking' the Norwegian public purse for as much as possible.

The 'Cold Peace discourse' is found also in matters related to nuclear safety in the European Arctic, but is less prevalent than in the management of marine living resources in the Barents Sea. Primarily, this discourse serves to obscure Norwegian motivations in the eyes of many Russians: why should Norway spend large sums of money to help solve local nuclear safety problems in Northwestern Russia if the real objective is not somehow to destroy Russia? The closure of the Kola nuclear power plant, required by the Norwegian side, can also be viewed from this angle: such a closure would be a serious blow to the local population and regional economy, a situation which would – allegedly – serve to secure Norwegian interests ('the loss of one state is always to the gain of other states').

Discourses on industrial pollution

The two nickel smelters on the Kola Peninsula – the Pechenganickel Combine at Zapolyarnyy/Nikel and the Severonickel Combine at Monchegorsk – emit large quantities of sulphur dioxide (SO_2), causing considerable acid precipitation on the Kola Peninsula and in the neighbouring Nordic countries. Emissions decreased during the 1990s as the result of a general slump in industrial output, but that considerable areas around the smelter towns have been irreversibly damaged by the pollution is incontestable. While current emission levels are in accordance with Russia's international obligations, industrial pollution from the Kola Peninsula nickel smelters is still considered by Western governments to be a major environmental challenge in the European Arctic. Large modernisation projects have been planned for the smelters from the Nordic side since the mid-1980s. It was not until 2001 that agreement was reached between Norway and Russia on a project that will involve a 90 per cent reduction in emissions of SO_2 and heavy metals, scheduled to be finalised in 2006–2007.

Two new discourses were defined in this case study: the 'death clouds discourse' in Norway and the 'anti-hysteria discourse' in Russia. In addition, two of the discourses found in relation to nuclear safety were encountered also here: the 'Barents euphoria discourse' combined with the 'sustainability discourse' on the Nordic side, and the 'environmental blackmail discourse', found on both sides of the border. The issue of industrial pollution from the Kola Peninsula nickel smelters features, above all, in the interface of the Nordic 'death clouds discourse' and the Russian 'anti-hysteria discourse'. The Kola Peninsula is, as far as the Nordic public is concerned, a catastrophe zone, a 'moon landscape', 'nuclear war zone' or 'black desert' – even small children 'know' that visiting Kola can make you ill! Although Russians generally acknowledge that the nickel smelters cause considerable pollution, they tend to give it a lower place in their 'hierarchy of problems' than their Nordic neighbours. First, they emphasise that the Kola Peninsula is for the most part a pristine corner of the world, blessed with many natural assets and the purest water and air. Second, they tend not to 'go all hysterical' about the pollution that does exist. They admit that it is a bad thing, but point to the fact that there are many other bad things in life and, moreover, new problems would be created if the smelters were forced to close down. The 'Barents euphoria discourse' of the early and mid-1990s – combined with the 'sustainability discourse' permeating the entire BEAR venture – resulted in something of a song and dance about 'our common future' and so on. The euphoria has faded, but while the dancers and singers still dance and sing, it is no longer about the environment. The Russians have finally managed to devise a cleaning scheme for the Pechenganickel smelter that has convinced the Norwegians to pay out the larger part of the money they had set aside for modernisation purposes as early as 1990. However, the early enthusiasm for this large enterprise is now largely replaced by the idea that the money will go to the 'limousine-driving stock holders' of Norilsk Nickel, and that as well running a company as Norilsk Nickel should be made to pay for its own modernisation.

Brokering scientific knowledge

Managing marine living resources, nuclear safety and industrial pollution are all 'trans-scientific problems', i.e. problems that require scientific knowledge, but cannot be solved by science alone (Litfin 1994; see also Chapter 1 in this book). The application of scientific knowledge varies across the three case studies presented in this book. In nuclear safety and industrial pollution, science played a part primarily by confirming that there was indeed a problem or, alternatively, that the problem in question was not as grave as anticipated. For instance, the scientific investigations carried out under the auspices of the AEPS/AMAP (see Chapter 2) confirmed the extent of forest death, as well as less serious environmental degradation, as a consequence of the sulphur emissions from the nickel smelters on the Kola Peninsula. Likewise, several scientific projects under the Norwegian Plan of Action for nuclear safety in Northwestern Russia stated the current level of radiation at sites where radioactive waste or spent nuclear fuel was stored. On the other hand, the joint Russian–Norwegian scientific expeditions to the Barents and

Kara Seas to investigate radiation levels in areas where the Russians had dumped radioactive material (see Chapter 4), served to reassure those who had feared nuclear pollution in the marine environment. These scientific data were 'brokered' to some extent in the ensuring process: Norwegian and Russian authorities used data from the Barents and Kara Seas expeditions in their efforts to convince the market that Barents Sea fish was 'clean'. It was further decided to leave the submerged material where it was (finding in the scientific data support for the view that costly efforts to raise the material were unnecessary). Data about industrial pollution were, for their part, 'brokered' by finding that the planned modernisation of the Pechenganickel smelter was indeed justified. However, scientific studies were not really necessary in this case; everyone could see that large areas around the nickel smelters had suffered from the emissions. Moreover, the decisions to launch massive assistance programmes from Norway in both the nuclear safety and industrial pollution sectors can be explained more by other factors than scientific knowledge; compare the discussion later on story lines and metaphors, as well as the embeddedness of discourses on the environment in more overarching discourses in society.

The continuous production and interpretation of scientific knowledge is far more prevalent in the management of marine living resources in the Barents Sea. This follows from the nature of the problem: the Joint Fisheries Commission sets annual catch quotas on the basis of recommendations from the scientific organisation ICES. It seems obvious that scientific knowledge – based here on something very close to complete agreement among scientists – was not crucial in determining the level of catch quotas; the TAC for Northeast Arctic cod was quite systematically set far above the scientific recommendations around the turn of the century. Moreover, the data were clearly 'brokered' by various interest groups in Norway and Russia and hence served to ignite both active and dormant discourses. For the Norwegian coastal fishing fleet, environmental NGOs and parts of the political left, the ICES recommendations were seen as implying that the trawler fleet was far too effective and probably also engaging in massive violations of the fishery regulations. This activated the critical variant of the 'sustainability discourse'. For many fishers in both Russia and Norway, the pessimistic scientific assessments were simply 'proof' that marine science is totally out of sync with the actual situation at sea, compare the assumptions of the 'seafaring community discourse'. Likewise, large segments of the Northwest Russian population saw in ICES's recommendations nothing but yet another indication that international organisations dominated by the West were ready to harm Russia, thus igniting the 'Cold Peace discourse'. As in the cases of nuclear safety and industrial pollution, however, the effects of scientific knowledge on the management of the Barents Sea fish stocks seem to be closely linked to the embeddedness of case-specific discourses in more general discourses. Moreover, the embeddedness of discourses on marine living resources in other discourses propels conflicts of interest and diverging views on institutional arrangement on the basis of how the scientific knowledge is 'brokered'. How scientific data are interpreted also affects how interests and institutional concerns are framed. The prevalent Russian 'Cold Peace' discourse, which views ICES recommendations as more or less dictated by powerful Western governments, also sees the 'quota distribution game' as a battle between the states involved, i.e. Norway and Russia. Consequently, the Joint Fisheries Commission is discredited as the major international institutional arrangement in the area's fisheries management. In Norway, the placement of subjects in either the 'official' or 'critical' versions of the 'sustainability discourse' as a result of how scientific data are brokered, also determines one's views in questions of interests and institutions. Or the other way around.

Story lines and metaphors

As we saw in Chapter 1, Hajer (1995: 56) defines a story line as 'a generative sort of narrative that allows actors to draw upon various discursive categories to give meaning to specific physical or social phenomena'. Dryzek (1997) discusses how various discourses employ different metaphors and other rhetorical devices. Story lines and metaphors have quite similar functions in a discourse, catching certain aspects of a problem in a simple and understandable manner and establishing 'the way one speaks around here'. In my case studies, metaphors proved to have a strong effect in areas of nuclear safety and industrial pollution. Calling the Kola Peninsula a 'black desert', the clouds drifting from Russia into Scandinavia 'death clouds' and the Northwest Russian nuclear complex a 'ticking time bomb' – largely the achievement of the media, NGOs and environmentally oriented politicians – had a substantial effect on the general public in Norway and, in turn, its political establishment and government. Emerging in the late 1980s and the early 1990s, the 'nuclear disaster discourse' and the 'death clouds discourse', centring around the mentioned metaphors, were routinely reproduced as the established way of talking about the environment of the European Arctic, reflected even in children's plays on words. Other discourses, primarily the 'environmental blackmail discourse', have questioned the legitimacy of the Norwegian assistance schemes for Russia in the areas of nuclear safety and industrial pollution, but the premises of the 'nuclear disaster discourse' and the 'death clouds discourse' remained basically unquestioned – thanks, to a large extent, to the enormous effect of the 'disaster metaphors'. As an example from the Russian side, the metaphor of the Kola Peninsula as a pristine corner of the world supposedly had the effect of tempering the feeling that 'something needs to be done', expressed by the Norwegians.

During the discussion of major discourses in Chapters 3, 4 and 5, I defined the main story lines of each discourse. Not all of them had an equally strong impact within the discourse in question, and not all the discourses were of equal importance in the discourse genealogy of the subject matter. Moreover, some of the story lines had a strong influence in their own right on the overall discourse, inaugurating new, offshoot discourses in the field in question. Others I selected as examples of prevailing story lines within a discourse, though not necessarily as the statements that introduced a new discourse or changed the direction of an established one. The best example of the former is probably the announcement that Norway 'forced' the Joint Fisheries Commission to make in 1999: 'the difficult circumstances of the Northwest Russian people necessitate a high cod quota.' The announcement initiated the 'pity-the-Russians discourse' in matters of fisheries management in the Barents Sea. Until then, quota establishment had been an issue of sustainable/ precautionary management or lack of the same (in Norway), and of a battle between two coastal states (from the Russian point of view). Suddenly, fish quotas became a question of solidarity with a 'suffering' neighbouring people. The image of Northwest Russians as poor and deprived was easy to identify with for people in Norway and made sense of an otherwise illegitimate management practice. Although there is good reason to believe that the non-reduction in the cod quotas primarily served to make Russian *nouveaux riches* even richer, the story line activated an already well-established discourse in society in a new thematic area (fisheries management), and its premises were not even questioned.

Examples of story lines whose first appearance I cannot ascertain, and hence evaluate the immediate effects of, are those referred to in the Russian 'Cold Peace discourse', e.g., 'Of course, it's in Norway's interest to ruin Russia; this is simple economic theory.' I assume, nevertheless, that similar statements must have been made by various actors within the defined discourse coalitions which had the effect of 'clarifying' various phenomena – primarily the motivations of the Norwegians – that until then had remained incomprehensible for them. Somewhat differently, the story lines I referred to in the 'nuclear complex discourse' and the 'anti-hysteria discourse' – 'Our experts know what they are doing' and 'Yes, we do have a pollution problem, but what good does it do going all hysterical?' – reproduced a well-known narrative practice, i.e. gave meaning to what otherwise appeared as incomprehensible Norwegian 'hysteria'.

Embeddedness and discourse classification

A main conclusion of this study is that environmental discourses are embedded in more widespread social discourses. The examples above of knowledge brokerage and the effects of story lines and metaphors are mostly related to this embeddedness of discourses. Existing discourses are activated in new functional settings when knowledge is brokered or story lines or metaphors evoked. The best example is perhaps the clear embeddedness of the major Russian discourse on the Barents Sea fisheries management, the 'Cold Peace discourse', in a similar overarching discourse on relations between Russia and the West after the cessation of the Cold War. Another example is the incorporation, mentioned in the preceding section on the effect of metaphors and story lines, of the 'pity-the-Russians discourse' into the complex of discourses on marine living resources in the Barents Sea. Since the early 1990s, the Norwegians had been bombarded, by the media, politicians, NGOs and others, with stories of calamities and suffering in Northwestern Russia. The 'pity-the-Russians discourse' was there, ready to be activated in a new functional field. The influence of the general 'Barents euphoria discourse' on environmental discourses is more indirect; it developed parallel to the 'nuclear disaster discourse' and the 'death clouds discourse' and fuelled enthusiasm to help out the Northwest Russian population.

Table 6.1 lists the main discourses as defined in the three case studies. Instead of Dryzek's classification into 'industrial' and 'environmental' discourses, I use the terms 'eco-centric' (Dryzek's 'environmental') and 'techno-centric' (Dryzek's 'industrial') discourses.[2] The overarching discourses in society are, hence, labelled 'non-environmental' (i.e. not primarily oriented towards matters of the environment), and those specifically directed at various aspects of environmental management are named 'environmental'. As expected, eco-centric discourses predominate in the West; techno-centric ones are mainly found on the Russian side. The 'sustainability discourse', with its emphasis on concepts such as 'sustainable development' and the 'precautionary principle', is the typical eco-centric discourse among those singled out in the three case studies. It resembles Dryzek's archetypical sustainability discourse, which foresees imaginative attempts at solving the conflicts between environmental and economic concerns, but only in a reformist way. The 'nuclear disaster discourse' and 'death clouds discourse' are less based on a scientific or theoretical-principal fundament. They reflect more the ideas and values of eco-centric discourses in general. Probably, it would be more correct to understand them within the frame of what Dryzek calls 'green radicalism', seeking imaginative departures from the status quo in a radical way.

The three Russian discourses grouped together under the heading 'techno-centric discourses' in Table 6.1 – the 'seafaring community discourse' (admittedly also found on the Norwegian side), the 'nuclear complex discourse' and the 'anti-hysteria discourse' – are all variations over the same theme. They reflect the typical Soviet-type belief in economic and social progress through the conquering of the natural environment, and a corresponding neglect of possible negative consequences of industrial activity. The fishing fleet, nickel smelters and the nuclear power plant on the Kola Peninsula are symbols of (Soviet) man's conquest of the natural world, bringing food, electricity, employment and economic gain to the local community, the region and the Union (today: Motherland). The labelling of the three discourses as the 'seafaring community discourse', 'nuclear complex discourse' and 'anti-hysteria discourse' is somewhat haphazard, or at least dependent on the concrete and time-specific situation depicted here.

Table 6.1 Main discourses on the environment in the European Arctic during the 1990s

Non-environmental	• 'pity-the-Russians discourse' (Norway/marine living resources)
	• 'Cold Peace discourse' (Russia/marine living resources)
	• 'Barents euphoria discourse' (Norway/nuclear safety and industrial pollution)
	• 'environmental blackmail discourse' (Norway and Russia/ nuclear safety and industrial pollution)

Environmental	*Eco-centric discourses*	*Techno-centric discourses*
	• 'sustainability discourse' (Norway/marine living resources and industrial pollution)	• 'seafaring community discourse' (Norway and Russia/marine living resources)
	• 'nuclear disaster discourse' (Norway/nuclear safety)	• 'nuclear complex discourse' (Russia/nuclear safety)
	• 'death clouds discourse' (Norway/industrial pollution)	• 'anti-hysteria discourse' (Russia/industrial pollution)

One could easily have labelled all three of them variations over the 'industrial complex discourse'. A 'fisheries complex discourse' is discernible within the somewhat wider 'seafaring community discourse', which also embraces 'fish folk' beliefs and values among Norwegian fishers and departs from the general 'industrial complex discourse' in its time-specific scepticism towards scientific knowledge.[3] The 'anti-hysteria discourse' related to industrial pollution, in turn, resembles what could have been referred to as the 'nickel complex discourse'. In order to retain certain nuances of the fisheries discourse (notably the 'we know best' argument and the sense of fellowship between Norwegian and Russian groups of fishers in this situation) and the discourse on industrial pollution (the lack of 'hysteria' compared to the Norwegian discourse), I have chosen to emphasise slightly different aspects of the three discourses. With a few modifications, however, they can all be regarded as variations of a more general 'industry complex discourse'.

Some of the Norwegian discourses are also variations of the same theme. Notably, the 'nuclear disaster discourse' and the 'death clouds discourse', albeit here referred to as 'environmental' discourses due to their pre-occupation with specific environmental matters (see Table 6.1), can be understood as slight modifications of a more encompassing 'pity-the-Russians discourse'. Northwestern Russia has generally been spoken about in terms of calamities and suffering in Norwegian society since the late 1980s, and this has crystallised as the established way of talking also in matters of the environment. As mentioned in Chapter 3, Norwegian journalists have difficulties selling other types of stories about the Kola

Peninsula than those relating to the Kursk accident, the 'ticking time bombs' of the nuclear complex, environmental degradation and suffering communities. If the 'industry complex discourse' can encapsulate the most important aspect of Russian environmental discourse related to the European Arctic since the late 1980s, the prevailing discourse on the Norwegian side has been an overarching 'pity-the-Russians discourse', or, perhaps, rather 'pity-the-Russians (and-pity-us-who-live-so-close-to-them) discourse'.

The environment, Russia and the West

Can any lessons of a more general nature be extracted from the case studies in this book? In addition to the theoretical remarks made in the preceding sections about the interconnectedness of knowledge brokerage, employment of story lines and metaphors and the embeddedness of discourses, a few words about the East–West interface in environmental affairs seem appropriate. First, and not very unexpectedly, the Russian discourse seems rooted in a techno-centric metadiscourse, and the Norwegian in an eco-centric metadiscourse – at least in their modes of talk. As we saw in Chapter 3, the Norwegians maintained an ear-splitting 'sustainability chorus' in the management of the Barents Sea fisheries, even after the quota for several years had been established at a level far above what scientists considered to be in accordance with the precautionary principle. In the BEAR collaboration, admittedly incorporating the regions of four countries but initiated and led by Norway during its first years, the 'sustainability chorus' permeated the entire venture with its constant chants of 'sustainable development' – even puppet theatres were given money to spread the good word! The Russians, for their part, seem mainly to act within an 'industry complex discourse', emphasising the positive effects of the fishing fleet, nuclear complex and nickel smelters of the area. These are viewed as contributing primarily to development and wealth in the region, rather than to the degradation of the environment and health.

Second, the leading discourses on the Norwegian side, at least within the areas of nuclear safety and industrial pollution, seem to be founded on 'disaster metaphors' more than on scientific facts and internationally acknowledged principles.[4] As elaborated in Chapters 4 and 5, the Norwegian 'nuclear disaster discourse' and 'death clouds discourse' clash fundamentally with the nuclear and nickel variants of the Russian 'industry complex discourse', which in turn puts a premium on expert opinion and scientific knowledge. The close links between 'media hysteria' and political action in Norway – politicians obviously have to 'do something' when demanded by public sentiment – lead to what from the Russian side must sometimes appear as 'tabloid politics'.

A few general remarks are in their place at this point. Some of the typical Western 'modes of talking' about the environment are not as widespread in Russia as in the West. Techno-centric discourse, largely a relic of the past in the West, is still prevalent in Russia. Moreover, the 'hysteria' of Western 'tabloid politics' founded on 'disaster metaphors' is often not taken seriously by leading players in Russia – 'This is an issue for our experts, not for the *narod* (people)!' When Western politicians, nevertheless, steadily pursue politics filled with 'sustainability slogans' – the Murmansk-based newspaper *Polyarnaya Pravda* once talked of 'fanatic democrats' – or keep referring to the need to do something to ease the nervous *narod*, many Russians tend to become insecure about the real motivations of Western governments. This, in turn, activates familiar categories depicting the West as interested in little else than destroying Russia. The story of quota establishments in the Barents Sea from 1999 is telling: the establishment of cod TACs is certainly *not* an issue of sustainability in the eyes of the Russians, such actions simply fuel suspicions of a hidden agenda in the Norwegian stance. Likewise, the Russians consider the motives behind some of the projects under the Norwegian Plan of Action for nuclear safety in Northwestern Russia as distinctly unclear. In both cases, the dominant Russian 'Cold Peace discourse' provides ready answers: the Norwegians are affluent enough to abstain from fishing quotas only if doing so means dealing the final blow to the Russian fishing industry; their 'good neighbours' are ready to donate new infrastructure for the Kola nuclear power plant,[5] if the Russians only promise to shut down according to schedule and hence strangle a local community and the energy supplies for an entire region; the loss of one state is always to the benefit of other states.

The aim of this book has been to investigate how the way we talk about the environment influences politics in the field, not to evaluate the results of these politics. Hence, I am not giving recommendations as to how the West should organise its relations with Russia in environmental policy – or the other way around.[6] The Nordic (mainly Norwegian) experience recounted here might, nevertheless, have some relevance for other Western governments involved in environmental co-operation with Russia. First and foremost, variants of the prevailing Western eco-centric discourse do not always fit well with the Russian (techno-centric) 'industry complex discourses'. What people in the West see as stirring metaphors and timely insistence on important principles, is often seen by the Russians as examples of environmental 'hysteria' and 'fanaticism'. With the Russians so strongly positioned in the 'Cold Peace discourse' – or in a more timeless higher-order discourse whose main premise is that states are persistently in conflict with each other – Western motivations tend to be interpreted in the worst possible way. This tendency escalates further if Russians feel that Western motives are unclear or simply 'too altruistic to be true'. On the other hand, one could speculate whether it would be easier for Russians to take advantage of Western goodwill – in line with the 'environmental blackmail discourse' – if they thought there was a fair chance that the ultimate Western goal – concealed behind smoke screens of good intentions – was nevertheless to ruin Russia. In that case, one recommendation for Western governments would hence be to make their intentions as explicit as possible for the Russians.

Further, Western motives would need to be 'translated' into Russian preconceptions to enhance understanding. In particular, one should be aware that environmental 'hysteria', 'euphoria' and other forms of 'fanatical democracy' are not easily 'bought' by the Russians; pure national self-interest is, however, a language understood by people situated in the state-centric metadiscourse. Western assistance to Russia should, perhaps, include introducing social sciences that depict international relations as more than 'simple economic theory'.

Notes

1 The study of environmental discourse

1 It is also one of several ways in which I have approached the subject matter of the book. It is my conviction that discourse analysis can be a *supplement* to more traditional political science approaches to the environmental interface between Russia and the West in the Arctic, not a *replacement*.

2 Management of natural resources and the environment is often grouped together in policy studies and commonly labelled 'environmental studies'. For the sake of consistency and simplicity, I will use the term 'environment' in this book to cover both natural resources and the rest of the natural environment.

3 The 'European Arctic' is here understood as the eastern part of those parts of the European continent located north of the Arctic Circle. This part of the European Arctic is necessarily of most relevance as long as focus is on Western interaction with Russia. In practice, the geographical scope of the book encompasses the northernmost parts of Norway, Sweden, Finland and European Russia. Included in the concept is also the Barents Sea, i.e. those parts of the Arctic Ocean lying between the North Cape of the Norwegian mainland, the South Cape of the Spitzbergen Island of the Svalbard Archipelago and the Russian archipelagos of Novaya Zemlya and Franz Josef Land. In the discussion of the interface between Russia and the West in environmental matters, the parts of the region located closest to the East–West border, i.e. Murmansk Oblast in Russia and Norway, as well as the ocean areas of the Barents Sea, play the most important role.

4 See, e.g., Hønneland and Blakkisrud (2001).

5 For brief overviews of the history of the Kola Peninsula, see, e.g., Hønneland and Jørgensen (1999a, 1999b) or Hønneland and Blakkisrud (2001).

6 Ibid.

7 For an overview of Northwest Russian fisheries, see Hønneland (1998a, 2000a).

8 See AMAP (1997, 1998) for reviews of the state of the environment in the Russian Arctic.

9 Previous schemes were finally buried in 1997, but a new agreement on the modernisation of the Pechenganickel smelter was concluded between Norway and Russia in June 2001. The new project is scheduled to be completed in 2006–2007.

10 The Chernobyl accident in April 1986 was, of course, the 'big awakening' for the European public to the dangers of nuclear radiation. The *Komsomolets* accident, for its part, served to remind the public that radiation can emanate from other sources than power plants, still by far the most serious threats to the general public.

11 See, e.g., Stokke (1998, 2000a).

12 See Nilsen *et al.* (1996) for an assessment of the threats of radioactive contamination emanating from the Russian Northern Fleet.

13 As mentioned above, the extraction of stationary resources on the Kola Peninsula has decreased in recent years due to economic problems at company level.

14 See, e.g., Litfin (1994), Hajer (1995), Tennberg (2000) and Neumann (2001).

15 Ibid.

16 Litfin (1994) contrasts the reflectivist approach with those of neorealism and neoliberal institutionalism, both of which, she argues, are objectivist, taking goals and interests as given. The reflectivist approach, while also preoccupied with the study of social structures, insists that structures are made up of identities and interests and cannot exist apart from processes.

17 Litfin (1994) refers to Goldstein (1989), Hall (1989) and Kratochwil (1990).

18 See, e.g., Hajer's (1995) reference to Kuhn (1962/1970), Berger and Luckmann (1966/1984), Giddens (1979, 1984) and Douglas (1982, 1987). The essence of the social constructivist conception of politics is, in rough terms, that some issues are organised into politics while others are organised out (Schattschneider 1960; Torgerson 1990).

19 See, e.g., Hajer's (1995) reference to Bernstein (1976/1979) and Gibbons (1987). The main argument of the interpretative approach is that general laws and causality characteristic of natural science cannot be found in the social sciences, and that the latter should instead aim at tracing conceptual connections and elucidating the meaning of social processes.

20 Neumann (2001: 26) emphasises the affinity between the critical Frankfurt school and discourse analysis in that they both insist on cross-field approaches, have as their goal to produce new readings of social life, are preoccupied with the relationship between power and knowledge and look into how social orders are produced and function.

21 Compare Barth (1993).

22 Compare Bartelson (1995).

23 Compare Foucault (1972).

24 As we have seen, the statements are themselves sometimes perceived of as the contents of the discourse.

25 I am clearly more comfortable with definitions that portray discourses as a 'process' or 'practice' rather than as a 'system' or – as some maintain – as the products (concepts, ideas etc.) themselves.

26 For instance, Wissenburg *et al.*'s (1999) collection of essays entitled *European Discourses on Environmental Policy* consists mainly of 'mainstream' social science research on environmental politics with sporadic allusions to the theoretical discourse debate. The same can be said about Eder and Kousis (2000), which carries the subtitle *Actors, Institutions and Discourses in a Europeanizing Society*. Cocklin (1995) is an example of the use of the term in works that basically review various traditions in a given research field.

27 Other related studies include Cantrill and Oravec's (1996) case studies of 'environmental communication', Lash *et al.*'s (1996) largely post-structuralist study of risk, environment and modernity, Myerson and Rydin's (1996) investigation of environmental rhetoric and Tennberg's (2000) discourse analysis of the establishment of the Arctic Council.

28 A brief overview of the main theoretical and empirical arguments of this work is provided in Litfin (1995).

29 During Litfin's (1994: 6) own empirical investigation, '[i]t became increasingly evident that "knowledge" was not simply a body of concrete and objective facts but that accepted knowledge was deeply implicated in questions of framing and interpretation and that these were related to perceived interests.'

30

'[A] discursive approach should not pretend that social agents are either nonexistent or unimportant, despite the language of some poststructuralists.

Without agents promoting them, identifying with them, and struggling over them, discourses could not exist; but agents do not act autonomously, wielding the power of discourse on behalf of transparent interests.

(Litfin 1995: 253)

31 Interestingly, Litfin and Hajer have no reference to each other's works.

32 Hajer accentuates acid rain as a particularly suitable subject for discourse examination since it is 'beset with uncertainties. . . . [A]s a creeping and cumulative form of pollution acid rain was a typical example of the new generation of environmental issues that, more than their predecessors, depend on their discursive creation' (Hajer 1995: 6). Acid rain is a multidimensional problem with many possible solutions, illustrating the importance of 'problem framing' for political measures.

33 Still, he finds both Litfin's and Hajer's works 'productive' (Dryzek 1997: 9).

34 He quotes Litfin's (1994: 26–27, 50) claim that 'it is possible to subscribe to both a hermeneutic epistemology (i.e. an interpretive philosophy of inquiry) and a realist ontology (i.e. a commitment to the actual existence of problems)' (Dryzek 1997: 10).

35 The organisation of each case study varies somewhat; while the chapter on marine living resources has specific sections about discourses on knowledge, interests and institutions, these issues are more indirectly treated in the case studies on nuclear safety and industrial pollution.

36 Hence, the term 'institution' is used in a fairly broad sense here, embracing all routinised practices between Russian and Western partners in the given domain of politics. In practice, most of these forms of organisation could also be characterised as international regimes, understood as 'sets of implicit or explicit principles, norms, rules, and decision-making procedures around which actors' expectations converge in a given areas of international relations' (Krasner 1982: 186).

37 The research projects range from investigations of primarily a theoretical nature to empirically oriented studies, evaluations and consultancies for public authorities, industries and NGOs.

38 My interviewees include representatives of public authorities at both national and regional level in Norway and Russia, of industries causing the environmental problems, NGOs, scientific institutes, target groups (e.g. fishers) and others.

39 I was an interpreter for the Norwegian Coast Guard in the Barents Sea from 1988 to 1993. Since then, I have taken on occasional assignments as an interpreter, particularly for Norwegian fisheries management authorities. Until 2000, I was the interpreter of the Permanent Committee on Fisheries Regulation and Enforcement under the Joint Norwegian–Russian Fisheries Commission. On one occasion, in 1999, I also interpreted at the annual session of the Joint Commission itself. Reference is made only to events that have already been made public by people involved in these management activities.

40 During the second half of the 1990s, I usually spent a week in Russia, mainly Murmansk and Moscow, on average every other month. These visits have given me a good grip not only of my objects of research, but also of general trends – or discourses – in Russian society.

41 Murmansk is one of the Russian cities where a Lenin statue still overlooks the main avenue – the Lenin Prospekt.

42 Former Northern Fleet officer Aleksandr Nikitin was accused of espionage while collecting data on the nuclear risk on the Kola Peninsula for the Norwegian environmental NGO Bellona. Charges were finally dropped in 2000.

43 Compare a news bulletin from Radio Free Europe/Radio Liberty under the heading 'Murmansk a Hotbed of Foreign Spies': 'Foreign intelligence services

have targeted Murmansk Oblast as a "priority" area for their activities, Nikolai Zharkov, head of the Federal Security Service (FSB) directorate in Murmansk Oblast, told Interfax North-West on 28 December. . . . Zharkov also revealed that foreign governments frequently "pursue their own interests" under the cover of environmental organisations' (RFE/RL Newsline, 30 December 2000).

44 Even some 'official figures' agreed to an interview only after being introduced by mutual acquaintances.

45 Using a tape recorder has not been an option while conducting interviews in Russia since this would most certainly have made the interviewees far less likely to recount their most deeply held opinions and perceptions.

46 It could, of course, be objected that nor are the 'real' interview extracts word-for-word representation of what was said, especially as far as my interviews with Russians are concerned. One thing is that the interviewer necessarily misses small nuances (at best) since he or she is not always fast enough to note down in writing everything that is being said. Another is the fact that my interviews in Russia were conducted in Russian (by a native Norwegian!) and subsequently translated into English.

2 The environment and institutions in the European Arctic

1 See Stokke and Hoel (1991) for a discussion of how the quotas were shared between the two parties in Soviet times.

2 See Hønneland (2000b) for a discussion of the enforcement co-operation between Norway and Russia in the Barents Sea fisheries.

3 On the other hand, strategic submarines sometimes go to dismantlement well before the expiration of their service life.

4 The icebreaker fleet also includes conventionally powered ships.

5 However, even underground tests, both US and Soviet, have been known to cause radioactive fallout on the territory of other states (Stortinget 1994).

6 The Kola nuclear power plant and submarines at refuelling are both classified as sources of known or probable risks – i.e. 'release is known to have occurred or . . . significant probability for release has been confidently estimated' (Bergman and Baklanov 1998: 55). Sources of *potential* risk are 'those expected to constitute a risk for considerable release provided the outcome of further analysis for certain steps in the event chain' (ibid.).

7 Russia has signed, but not yet ratified START-2.

8 Reviews of the AMEC co-operation are given in Sawhill (2000) and Sawhill and Jørgensen (2001).

9 Admittedly, the current draft agreement is stricter than the Russian–Norwegian Framework Agreement as it includes provisions on personnel immunity, access and oversight.

10 The *urban-type settlements* of Nikel and Zapolyarnyy are the largest settlements of Pechenga *rayon* (district). There is also an *urban-type settlement* named Pechenga belonging to the *rayon* with the same name; this is considerably smaller than Nikel and Zapolyarnyy. For the sake of simplicity, the two latter are here referred to as towns (Zapolyarnyy is situated close to Nikel).

11 RAO is *Rossiyskoye aktsionernoye obshchestvo* (Russian stock-holding company).

12 No global convention exists on land-based pollution control; the main initiatives to date have shunned a strict precautionary approach to pollution control (VanderZwaag 2000).

13 See, e.g., Hanf (2000) for an overview of how the acid rain regime evolved.

14 The Regional Council is composed of representatives of the three northernmost counties of Norway, the two northernmost counties of Sweden and Finland, and

Murmansk and Arkhangelsk Oblasts, the Republics of Karelia and Komi and Nenets Autonomous Okrug in Russia, as well as a representative of the indigenous people of the region, the Saami. At the national level, the Barents Council consists of government representatives from Russia, the five Nordic countries and the European Commission. For discussions of the establishment and performance of BEAR, see Dellenbrant and Olsson (1994), Stokke and Tunander (1994), Dahlström *et al.* (1995), Dellenbrant and Wiberg (1997) and Flikke (1998).

3 Discourses on marine living resources

1 The major share of empirical data – mainly in the form of interview transcripts and extracts from the media and official documents – is presented in the middle section, which outlines discourses on knowledge, interests and institutions. However, certain new elements are introduced in the final section, defining the major discourses on marine living resources.

2 The precautionary *approach* is considered to be a less stringent variant of the precautionary *principle*. The principle originated towards the end of the 1980s in international law on pollution, but was accepted in international law on fisheries only after initial suspicion that it would mean the frequent use of ban on fisheries had been overcome. The use of the term *approach* is supposed to indicate a more flexible application of precautionary measures than the principle.

3 A common-pool resource can be understood as a natural (or in some cases synthetic) resource sufficiently large to make it costly to exclude users from obtaining subtractable resource units. Hence, the two criteria used to define a common-pool resource are first, the cost of achieving physical exclusion from the resource, and second, the presence of subtractable resource units (Gardner *et al.* 1994). The latter entails resource units acquired by one user taking place at the expense of other potential users. For example, this is the case with fish, but not with weather forecasts. Common-pool resources are distinct from public goods inasmuch as subtractability is low in the latter, and from private goods to the extent that exclusion is more difficult (Ostrom *et al.* 1994).

4 The introduction of the precautionary principle in global fisheries management is discussed in Garcia (1994), Hewison (1996) and Kaye (2001). A wider discussion of the implementation of the principle in various areas of international law is found in Freestone and Hey (1996).

5 The basis for calculating reference points is detailed further in a second report from the Study Group (ICES 1998a).

6 Although the phrasing 'resource management' is used, in practice it is fisheries management that is referred to; the passage is taken from a White Paper on Norwegian fisheries policy.

7 A fisheries act has been under preparation in the Federal Parliament since 1993. See Hønneland and Jørgensen (2003) for a discussion of this process.

8 Interview with Norwegian fisher, Båtsfjord, June 1998.

9 Referring to repeated refusals by Russian authorities in 1997 and 1998 to applications from Norwegian and Russian marine scientists to mount joint scientific expeditions in the Russian zone of the Barents Sea.

10 Interview with Norwegian fisher, Båtsfjord, August 1998.

11 Interview with Russian fisher, Båtsfjord, November 1997.

12 Interview with Russian fisher, Båtsfjord, August 1998.

13 The two investigations are not directly comparable. The compliance study involved interviews with individual fishers, i.e. captains of fishing vessels, while it was primarily shipowners who voiced their opinions in the debate about quota levels around the turn of the millennium. Further, the former investigation was

geared towards revealing general attitudes while the second concentrated on attitudes to scientific advice on quota levels. Lastly, the first study was done in 1997–1998 when stocks were relatively abundant, the second at the turn of the century when stocks had fallen again.

14 The views on scientific advice found among Norwegian fishers are more divergent than those found among Russian fishers. One reason might be that the Norwegian fishing fleet in less uniform than the Russian one and the various groups 'use' the issue of scientific advice to propagate the special interests of their own group; compare the discussion later in the chapter on discourses on interest in Norway.

15 *Fiskeribladet*, 23 November 1999, p. 3.

16 Russian shipowner to *Fiskeribladet*, 15 September 2000, p. 20.

17 *Fiskeribladet*, 14 December 2001, p. 6.

18 Announcement by a regional branch of the Norwegian Association of Fishers published in *Fiskeribladet*, 19 November 1999, p. 5.

19 Norwegian fisher to *Nordlys*, 17 November 1999, p. 15 and *Fiskeribladet*, 19 November 1999, p. 13.

20 Leader of the Norwegian Association of Shipowners to *Fiskeribladet*, 14 January 2000, p. 14.

21 Leader of a regional branch of the Norwegian Association of Fishers to *Fiskeribladet*, 12 November 1999, p. 12.

22 *Nordlys*, 16 November 2000, p. 2.

23 *Fiskeribladet*, 9 November 1999, p. 4.

24 *Fiskeribladet*, 23 November 1999, p. 4.

25 *Fiskeribladet*, 23 November 1999, p. 2.

26 Deputy leader of the Norwegian Association of Fishers to *Fiskeribladet*, 12 November 1999, p. 8.

27 Norwegian marine scientist to *Fiskeribladet*, 12 November 1999, p. 6.

28 Norwegian marine scientist to *Fiskeribladet*, 18 May 2000, pp. 1–2.

29 Conversation with Russian scientist, Murmansk, October 2000.

30 Conversation with Norwegian scientist, Bodø, February 2002.

31 See *Fiskeribladet*, 5 October 2001, p. 2, for an interview with Russian marine scientist Vladimir Borisov on this issue.

32 Conversation with Norwegian marine scientist, Bodø, February 2002.

33 For an overview of the structure of the Norwegian fishing industry, see, e.g., Holm and Mazany (1995). For a presentation of the Norwegian fisheries management system, see Hoel *et al.* (1996).

34 Norwegian coastal fisher to *Fiskeribladet*, 16 November 1999, p. 10.

35 *Nordlys*, 13 November 1999, p. 8.

36 *Fiskeribladet*, 30 November 1999, p. 15. See also *Nordlys*, 13 November 1999, p. 10, for similar arguments promoting the interests of the indigenous coastal Saami population, and *Nordlys*, 25 November 1999, p. 2, where the north–south conflict in Norwegian fisheries is made more explicit.

37 Representative of the Northern Norway regional HQ of the environmental NGO Nature and Youth to *Fiskeribladet*, 30 November 1999, p. 14.

38 Of course, it could be argued that this is a natural consequence of a more uniform Russian fleet than the Norwegian one, consisting, as it does, nearly exclusively of trawlers. As follows from the discussion later in the chapter, however, it is my opinion that this is not the main reason for the direction that the Russian discourse has taken.

39 There have been no signs in official statements, in the press or in my numerous interviews with actors within the Russian fisheries sector of divergent opinions within the Russian delegation to the Joint Fisheries Commission. As far as the Russian population at large is concerned, it is naturally impossible to exclude

that there are people sympathetic to the Norwegian position, but I have been unable to find representatives of this view during my stays in Russia or in written material.

40 *Fiskeribladet*, 12 November 1999, p. 3. To Norwegian ears, this alludes to the nickname that former Minister of Fisheries Jan Henry T. Olsen was given in the early 1990s when the Norwegians were discussing whether to apply for EU membership or not: *no-fish-Olsen*. Minister Olsen had said at the time that Norway was not willing to give a single fish to the EU as part of a membership deal.

41 *Rybnaya stolitsa*, 15 November 1999, p. 1.

42 Conversation with Russian civil servant, Murmansk, February 2000.

43 *Rybatskiye novosti*, no. 3–4, 2001 (cited from an unpaginated copy of the article).

44 *Rybnyy biznes*, November 2000 (cited from an unpaginated copy of the article).

45 Leader of the Norwegian delegation to the Joint Fisheries Commission to *Fiskeribladet*, 17 November 2000, p. 2.

46 Representative of the Norwegian Ministry of Foreign Affairs to *Fiskeribladet*, 7 April 2000, p. 24.

47 *Rybatskiye novosti*, no. 3–4, 2001 (cited from an unpaginated copy of the article).

48 Conversation with inhabitant of Murmansk, April 2001.

49 *Fiskeribladet*, 14 December 1999, p. 4.

50 *Rybnyy biznes*, November 2000 (cited from an unpaginated copy of the article).

51 Vice Governor of Murmansk Oblast to *Murmanskiy vestnik*, November 2000 (cited from an unpaginated copy of the article).

52 Russian shipowner to *Fiskeribladet*, 12 October 1999, p. 3; referred from *Murmanskiy vestnik*, 23 September 1999.

53 *Nordlys*, 18 November 1999, p. 7.

54 IntraFish, http://www.intrafish.com, 2 November 2001.

55 See Singh and Saguirian (1993) for a discussion of the creation of the Svalbard regime.

56 See Ulfstein (1995) for a discussion of the legal status of the Svalbard Treaty. Churchill and Ulfstein (1992) provide an extensive discussion of the legal setting of the Barents Sea fisheries.

57 An exception relates to some third countries, for which the Joint Fisheries Commission may specify the share of their total quota that can be caught in the Svalbard area (the Commission does not refer to the Svalbard *Zone!*). The rationale for such a specification is not legal in nature. Rather, it follows from the fact that fish are bigger closer to the mainland. Hence, the coastal states prefer that as much as possible of third countries' catches are taken here in order to minimise the long-term harm to the stocks.

58 This can be explained by the fact that Norway allocates approximately 75 per cent of its cod quota in the Barents Sea to coastal vessels operating closer to the Norwegian mainland while the Russian fleet consists almost exclusively of ocean-going trawlers, compare the earlier discussion about conflicts of interest within the Norwegian fishing industry.

59 This happened for the first time in 1993, when Icelandic trawlers and Faeroe vessels under flags of convenience started fishing here. Warning shots were fired at the ships by the Norwegian Coast Guard, and the fishing vessels left the zone. The following year, an Icelandic fishing vessel was arrested for having fished in the Svalbard Zone without a Barents Sea quota.

60 Aftenposten Interaktiv, http://www.aftenposten.no, 22 April 2001.

61 CNN Norge, http://www.cnn.no, 25 April 2001. The Norwegian Coast Guard succeeded in 'catching' the net so as to secure evidence of the crime.

62 Aftenposten Interaktiv, http://www.aftenposten.no, 22 April 2001.
63 Norwegian authorities made no reference to the fact that this was by no means the first time that equally serious violations had been perpetrated by Russian trawlers in the Svalbard Zone. Before, the violators were given a written warning and not even fined.
64 Aftenposten Interaktiv, http://www.aftenposten.no, 23 April 2001.
65 Collaboration was resumed in June the same year, when the Joint Fisheries Commission convened for its annual 'follow-up' session, in the wake of the main session that usually takes place in November.
66 *Fiskeribladet*, 28 September 2001, p. 3.
67 *Nordlys*, 12 June 2001, p. 3.
68 When this was written, the Svalbard Treaty had actually been signed more than eighty years previously.
69 *Rybnaya stolitsa*, no. 39, 2000 (cited from an unpaginated copy of the article).
70 *Rybnaya stolitsa*, no. 24, 2001 (cited from an unpaginated copy of the article).
71 *Fiskeribladet*, 17 October 2000, p. 2.
72 *Fiskeribladet*, 9 November 1999, p. 6.
73 See, e.g., feature article by Norwegian Minister of Fisheries Otto Gregussen in *Fiskeribladet*, 7 September 2001, p. 13.
74 *Fiskeribladet*, 21 August 2001, p. 13.
75 *Nordlys*, 14 November 2001, p. 37.
76 *Fiskeribladet*, 13 November 2001, p. 4.
77 Most statements referred to here were made by environmental NGOs, regional political authorities and the coastal fishing fleet. For similar views expressed by the political parties on the left side, see, e.g., *Nordlys*, 26 June 2000, p. 17.
78 Leader of the Barents Sea office of the Norwegian Society for the Conservation of Nature to *Fiskeribladet*, 2 November 1999, p. 4.
79 Leader of the Barents Sea office of the Norwegian Society for the Conservation of Nature to *Fiskeribladet*, 2 November 2001, p. 7.
80 Editorial in the largest newspaper of Northern Norway, *Nordlys*, 7 June 2000, p. 2. The demand for a board of inquiry to investigate Norwegian fishery policies is repeated in an editorial of 8 November, p. 2.
81 Norwegian coastal fisher to *Dagbladet*, 7 November 2000, p. 4.
82 Open letter signed by politicians, fishers' representatives and artists in Northern Norway, as well as a former Minister of Fisheries. *Fiskeribladet*, 21 August 2001, p. 13.
83 Previous leader of the Barents Sea office of the Norwegian Society for the Conservation of Nature to *Nordlys*, 10 November 2000, p. 3.
84 *Fiskeribladet*, 7 November 2001, p. 4.
85 Of course, the agreements with Russia have not been beyond political control, but the Minister of Fisheries had until late 2001 met little resistance, or even interest, with regard to the agreements from parliament.
86 *Murmanskiy vestnik*, 18 September 1999, p. 3.
87 *Fiskeribladet*, 17 November 2000, p. 2. See also *Fiskeribladet*, 21 November 2000, p. 4, where Gusenkov refers to ICES as 'an instrument in the hands of the Norwegian government'.
88 The latter is actually a member of the former but often maintains its own positions in questions of fisheries management. On issues of quota settlement, the positions of the two associations have generally concurred in recent years.
89 Conversation with Russian fisheries researcher, Murmansk, February 2000.
90 A Western critique of the reforms inspired by the West is found in Cohen (2000), who refers to the events as a 'failed crusade'. Cohen coined the term 'Cold Peace' in an article first published in 1992. In a postscript to the article printed in Cohen (2000: 104), he says:

I do not claim a patent on the term *cold peace*, and am not even certain I was the first to use it in this context, but it subsequently began to appear frequently in the U.S. and Russian press – and remarkably even in a statement by Boris Yeltsin. (My article was published in Moscow, in Russian, under a similar title.)

91 *Polyarnaya pravda*, 10 March 1999, p. 2. See Hønneland and Jørgensen (1999a: 167–168) for a discussion of the revival of patriotic values in Northwestern Russia at the end of the 1990s.
92 For typical 'seafaring community' presentations of scientists and civil servants, see, e.g., *Fiskeribladet*, 14 December 2001, p. 6, where civil servants are criticised for seeing the world only 'from their nice offices in Oslo', and *Fiskeribladet*, 11 September 2001, p. 13, where the Norwegian Minister of Fisheries is said to have visited the fishing fields 'in patent-leather shoes'. Compare also the statement by the leader of the Norwegian Association of Coastal Fishers: 'for a period now, we must bear over with the cries of distress from *shipowners* and *bank directors* echoing in the mountains' (emphasis added).
93 It is extraordinary because protocols from sessions of the Joint Fisheries Commission usually contain only the conclusions of its discussions – e.g. the TACs agreed upon – and do not reveal anything about the positions of the two parties.
94 Some shops did in fact run out of some types of food because people had started hoarding in early September. This does not mean that they were emptied, however. Moreover, even though there was reduced availability, the situation stabilised in the course of a few weeks. As in the rest of Russia, large segments of the population were experiencing serious wage arrears towards the end of the 1990s, and many were hit by the increase in prices after the rouble's drop in value in mid-August 1998. Nevertheless, there were no signs of mass starvation on the Kola Peninsula. Even public institutions such as orphanages and hospitals continued to receive foodstuffs although they were not always able to pay for them.
95 Interview with representative of the regional committee co-ordinating receipt of humanitarian aid in Murmansk Oblast, Murmansk, September 1998.
96 It should be observed, however, that the large humanitarian organisations, such as the Red Cross and SOS Children's Villages, did a professional job based on thorough analyses of the actual needs of the people involved. Russian complaints mainly refer to the more spontaneous donations from towns and organisations in Norway. Here, foodstuffs, clothes and shoes were often collected without regard for actual requirements.
97 *Polyarnaya Pravda*, 23 September 1998, p. 1.
98 Ibid.
99 Interview in Murmansk, October 2001.
100 The other evil being 'non-agreement' with Russia, which would have challenged the legitimacy of the entire bilateral management regime of the Barents Sea fisheries.

4 Discourses on nuclear safety

1 For a recent discussion of relations between Northwestern Russia and Moscow, see Hønneland and Blakkisrud (2001). Jørgensen (2001) discusses these relations in the military sector in particular. Civil–military relations on the Kola Peninsula are discussed in Hønneland and Jørgensen (1999a).
2 Interview in Moscow, April 2000.

3 Interview with Russian civil servant, Moscow, April 2000.
4 See the section later in the chapter on project implementation.
5 Interview with Norwegian civil servant, Oslo, March 2000.
6 Interview with Russian civil servant, Murmansk, February 2000.
7 Interview with Russian civil servant, Murmansk, February 2000.
8 Interview with Russian civil servant, Moscow, April 2000.
9 See Chapter 2 for an overview of sums allotted to the Plan of Action and spent under it.
10 If not otherwise indicated, all information is based on the interviews. The figures on project economy are taken from the Ministry of Foreign Affairs (2000).
11 Interview with Norwegian civil servant, Oslo, March 2000.
12 Interview with Norwegian civil servant, Oslo, March 2000.
13 Interview with Russian civil servants, Murmansk, February 2000, and Moscow, April 2000.
14 American participation in AMEC projects is covered by the Nunn–Lugar Co-operative Threat Reduction (CTR) Programme framework agreement between the USA and Russia where those projects have a direct linkage to CTR activities (e.g. the elimination of ballistic missile submarines). AMEC projects without such a nexus are not covered by the CTR framework agreement.
15 Interview with Russian businessman, Murmansk, February 2000.
16 Interview with Russian civil servant, Moscow, April 2000.
17 See Hønneland and Moe (2000) for a further discussion of this.
18 Interview with Norwegian civil servant, Oslo, March 2000.
19 Interview with Russian civil servant, Moscow, April 2000.
20 Interview with Russian project participant and co-ordinator, Moscow, April 2000.
21 Interview with Norwegian project participant (Russian citizen), Oslo, March 2000.
22 Interview with Russian project participant and co-ordinator, Moscow, April 2000.
23 Interview with Russian project participant and co-ordinator, Moscow, April 2000.
24 Interview with Russian project participant and co-ordinator, Moscow, April 2000.
25 This was depicted by all our interviewees as a joint Norwegian–Russian initiative. However, it is hard to see that Russian concerns for markets for fish was as serious as the Norwegian since Russia at the time only exported a very limited part of its Barents Sea fish (Hønneland 1998a). Moreover, the Russian party asserts that it knew the areas in question were uncontaminated, but realised that Russian research reports would not be enough to convince people in the West.
26 As commented by two Norwegian professors in a newspaper feature article:

> The majority of people think radiation is dangerous even in very small doses. We are regularly amazed at studies showing that most people are obviously unaware of the fact that we live in an ocean of radiation. Radiation from radioactive sources in nature hits us twenty-four hours a day from cradle to grave.
>
> (*Aftenposten*, 3 September 2000; cited from an unpaginated copy of the article)

27 As one of the few critics, Norwegian researcher Edvard Stang in 1996 noted that: 'the Plan of Action covers areas that are of no environmental significance for Norway. We have no direct benefit from channelling money to secure dismantled submarines and stores of radioactive waste because the real threat comes from

the civilian nuclear power plant' (*Aftenposten*, 24 January 1996; cited from an unpaginated copy of the article). See Stang's views on the *Lepse* project on pp. 105–6.

28 I was myself also a victim of this generalised fear, although I was fully aware of the scientific facts that it was groundless: I once brought home from Murmansk a jar of home-made cranberry jam given to me by a Russian research colleague. Having a jar of jam from the Kola Peninsula in the refrigerator caused quite a stir among my friends and family. It was perceived as practically a contradiction in terms: cranberries from the Kola Peninsula are obviously radioactive. You just can't have radioactive material in your fridge. Shamefully, I must admit that the jam was never consumed.

29 *Aftenposten*, 24 January 1996 (cited from an unpaginated copy of the article).

30 In turn, the *Lepse* had initially been a symbolic issue for the Bellona Foundation. Here is how the initiation of the project was explained by Russian and Norwegian interviewees in the evaluation study of the Plan of Action: 'The initiative came from Bellona. Mr Ruksha, who was then technical director of Murmansk Shipping Company, was "knocking on every door" to get someone to take care of the problem, and Bellona reacted.' (Interview with Russian project participant, Murmansk, February 2000.) '*Lepse* became a kind of symbol of how bad things were. Bellona came up with several horrible suggestions as to how the problem could be solved. We simply had to include the project in the Plan of Action.' (Interview with Norwegian civil servant, Oslo, February 2000.)

31 'Region building' parallels 'nation building', but on an administratively lower level, i.e. conscious political attempts to establish the notion of a geographic area as a 'natural' entity in the minds of the population, compare Anderson's (1983) concept of 'imagined communities'. See Neumann (1994) for a discussion of post-Cold War region building in the European Arctic.

32 See Wiberg (1994) for a presentation of the hierarchy of conditions that would have to be met for the Barents region to develop into a functional region.

33 See Hønneland (1998c) for a discussion of the BEAR as an identity formation project.

34 In autumn 1994, I was invited by the Norwegian Ministry of Foreign Affairs to give a speech about 'the prospects for East–West co-operation in the Barents region' at a conference organised by the ministry. I suggested to discuss opportunities and barriers to such a co-operation, but was instructed to 'focus on the prospects since there are indeed great prospects'. The ministry representative seemed positively shocked that somebody could question the generally optimistic view of the day.

35 See Hønneland (1996) for a review and categorisation of the early social science literature on the BEAR.

36 Interview with Norwegian civil servant, Oslo, April 2000.

37 Interview with Norwegian civil servant, Oslo, March 2000.

38 Conversation with Russian researcher, Murmansk, February 2000.

39 Interview with Russian civil servant, Murmansk, February 2000.

40 It can be argued that the cut from the school at least shows an understanding for the possible negative consequences of having a nuclear power plant in the local community; the children have to be taught that it is not dangerous. After all, these events took place after the Chernobyl disaster.

41 At the time of writing, I am simultaneously working on an evaluation of the Barents Health Programme, covering health initiatives in Northwestern Russia financed by the Norwegian government. In some of the projects, the Norwegian project managers have tried – in accordance with the general Norwegian goal of encouraging the development of civil society in Northwestern Russia – to include Russian NGOs in joint projects with the Russian health sector, which tended to

be dominated by physicians (*mediki*). This was not always successful as some physicians tended to view the projects as a task for experts and did not see any need to bring in 'non-experts' (implicitly: charlatans). During an interview with an Arkhangelsk physician, our interviewee – referring to the interviewers' background in political science – several times complained that 'it is so difficult for me to discuss these things with non-*mediki*'. (Interview with Russian civil servant, Arkhangelsk, June 2002.)

42 Interview with Russian NGO activist, Murmansk, February 2000.

43 Of course, there is also an allusion here to the fact that it was a non-Russian environmental NGO that took the liberty to criticise the Russian nuclear complex.

44 Interview with Russian civil servant, Murmansk, February 2000.

45 A common theme in the economic development literature on the former Soviet republics – and the Russian Federation in particular – has been that regions located at the borders of the former Soviet Union (so-called gateway regions) may be better positioned for sustainable economic growth during the early post-Soviet period than regions in the interior. A key assumption underpinning this argument has been that a waning central governmental authority will allow regions more leeway to establish transborder economic ties with neighbouring states that are more developed, have abundant investment capital, or otherwise possess complementary economics to those of Russia's border regions (Kirkow 1997). For a discussion of the gateway region argument applied to Northwestern Russia, see Hønneland and Blakkisrud (2001).

46 This sentence is interesting for two reasons. First, it shows that the nuclear safety projects were in 1997 still being referred to as successful, as exceptions to the emerging rule that co-operation in the Barents regional was generally a 'failure'. Second, it exemplifies the prevailing Norwegian view that the Kola Peninsula was – contrary to scientific data – indeed 'nuclearly polluted' (compare the discussion of the 'nuclear disaster discourse' earlier in the chapter).

47 As follows later in the article, the author refers to the problems the Bellona Foundation encountered after the arrest of Aleksandr Nikitin (compare Chapter 1).

48 *Aftenposten*, 21 May 1997, p. 18.

49 Conversation with inhabitant of Murmansk, September 1998.

50 For evaluations of the projects within education and the health sector, see Hønneland and Moe (2002) and Jørgensen and Hønneland (2002).

51 *Aftenposten*, 24 January 1996 (cited from an unpaginated copy of the article).

52 Interview with Russian NGO activist, Murmansk, February 2000.

53 Our evaluation focused primarily on policy and organisational issues whereas the Auditor General's, naturally, was interested more in how the money had been spent.

54 *Aftenposten*, 24 January 2001 (cited from an unpaginated copy of the article).

55 *Aftenposten*, 11 January 2001, p. 4.

56 *Aftenposten*, 24 January (cited from an unpaginated copy of the article).

57 Conversation with Russian civil servant, Moscow, April 2000.

58 Such a massive loss of jobs and electricity would be contrary to the expressed Norwegian objectives for the area, namely to secure not only military and environmental stability, but also social stability in the area. However, the assumption seems to be widespread in Russia that every 'setback' sustained by the Russian population is simultaneously a victory for Western countries (compare the presentation of the 'Cold Peace discourse' in Chapter 3).

59 Admittedly, this logic does not conform with the logic of the 'environmental blackmail discourse', which says that the leaders of post-Soviet nuclear and polluting installations refuse to close the plant because it would mean the end of foreign financial assistance.

60 An overview of the Nikitin case is given at http://www.bellona.no/en/international/russia/status/4121.html.

61 RFE/RL Newsline, 30 December 2000.

62 Obviously, there are exceptions to this version: compare the story of how the owner of *Lepse* supposedly saw an economic advantage from keeping the population of Murmansk as 'disaster hostages'.

5 Discourses on industrial pollution

1 According to Kotov and Nikitina (1998b), the law came into force in 1980.

2 The Soviet State Hydrometeorological Service (Gidromet) was given the status of state committee and renamed the State Committee for Hydrometeorology and Environmental Monitoring (Goskomgidromet) in 1978, being elevated in the process from a lowly meteorological service to the most prominent environmental protection agency of the Soviet Union (Darst 2001). Today, the institution has the status of a 'federal service' and is usually referred to by the acronym Rosgidromet. This agency is no longer among the most important government bodies in Russian environmental management.

3 Goskomgidromet remained responsible for the monitoring of air pollution, but all other functions related to air-quality control were transferred to the State Committee for Environmental Protection.

4 Conversation with Norwegian researcher, Monchegorsk, November 2001.

5 *Natur & Miljø Bulletin*, no. 13, 1990.

6 Ibid.

7 Found at http://www.fi.uib.no/~btk/humor/html/alle.html as *Alle barna var friske unntatt Ola, han hadde vært på Kola.*

8 *Dag og tid*, no. 4, 1999. The review carries the heading 'Russia, a Neighbour in Deep Distress' and reflects an interesting mixture of the 'death clouds discourse' and the 'pity-the-Russians discourse' presented in Chapter 3. See Chapter 6 for a further discussion of how these discourses, as well as the 'nuclear disaster discourse', overlap.

9 See, e.g., http://odin.dep.no/odinarkiv/norsk/dep/shd/1997/pressem/030005–070193/index-dok000–b-n-a.html under the heading 'No Serious Health Problems from the Pollution on the Kola Peninsula'.

10 http://www.murman.ru/ecology/comitet.

11 Interview with Russian journalist, Murmansk, November 2001.

12 Another interesting observation made by several of the journalists interviewed, slightly outside the theme of this chapter, was that the Norwegians went on and on about the importance of a 'free press' and journalistic independence. But how much freer are Norwegian journalists, asked the Russians. They seem constantly to need 'disaster stories' from the Kola Peninsula (instead of more 'representative' stories about the region) in order to satisfy their editors, owners or readers.

13 See, e.g., Darst (2001) for a discussion of this.

14 There is one general 'depopulation of the North' programme; the others are for various groups such as pensioners, military personnel and others.

15 This point was also made by one of the candidates in Murmansk Oblast before the 1999 elections to the State Duma. 'My heart broke,' she said in a sad voice in a pre-election programme on a regional television channel, 'when I saw all the children who had to remain here in the north last summer'. One of her promises to the electorate was to put an end to this situation.

16 At one institution (notably for healthy, and not ill or disabled children), the director told us that close to half of the children were defined as 'sickly'. Asked which diseases predominated, she replied 'Arctic sight', i.e. reduced sight as a result of the long polar night! (Interview with Russian orphanage director, Kola, March 1998.)

17 See Ministry of Foreign Affairs (1993) for an overview of these areas.
18 In a twist of fate, this background document was written by me. As a social scientist at a research institute in Tromsø, Northern Norway, I was hired by the BEAR working group on cultural co-operation to draw up this document.
19 *Finnmarken*, 23 June 2001, p. 2.
20 This text was published in 1996. As stated early in this chapter, the State Committee on Environmental Protection was disbanded altogether in 2000, its remnants placed under the Ministry of Natural Resources.

6 Conclusion

1 Several Russian and Norwegian newspapers have been systematically read in search of articles on fisheries management, nuclear safety and industrial pollution issues. Most material was found on fisheries management, least on industrial pollution.
2 The terms are inspired by Benton and Short's (1999) 'ecological' and 'technological metadiscourse'.
3 I have elsewhere argued that the concept of a 'fishery complex' in Russia makes the Western distinction between 'authorities' and 'user groups' less applicable for studies of systems of fisheries management in post-Communist countries (Hønneland and Nilssen 2000; Hønneland and Jørgensen 2003). Whereas marine scientists, regulators, enforcers and the fishers themselves in the West are thought of as clearly distinguishable groups with their specific (and internally contradictory) interests, these categories of people tend in Russia to be perceived as a unity, all belonging to the 'fishery complex'. Again, the explanation lies probably in the vertical Soviet organisation of politics and industry, where entire sectors of the economy – like fisheries, atomic energy and nickel production – were treated as separate economic entities endowed with their own infrastructure.
4 This is not to say that the Bellona Foundation, for instance, does not have any scientific foundation for its recommendations. My point is that it is not the scientific facts, but the ingratiating metaphors and story lines that gave the 'nuclear disaster discourse' its magnitude and power.
5 The Northwest Russian press avails itself from time to time of the term 'our good neighbours' – in quotation marks – when referring to Norwegians, especially if the angle has something to do with perceived environmental 'hysteria' or humanitarian efforts coming out of Norway (the *gumanitarka*).
6 See Darst (2001: 210ff) for a reasonable list of such recommendations.

Bibliography

Aasjord, B. (2001) 'Norsk-russisk rulett i Barentshavet? Fiskeriforvaltning i lys av havrett og internasjonal folkeskikk', *Internasjonal politikk* 59: 303–332.

AMAP (1997) *Arctic Pollution Issues: A State of the Arctic Environment Report*, Oslo: Arctic Monitoring and Assessment Programme.

AMAP (1998) *Arctic Pollution Issues: A State of the Arctic Environment Report*, Oslo: Arctic Monitoring and Assessment Programme.

Anderson, B. (1983) *Imagined Communitites: Reflections on the Origin and Spread of Nationalism*, London: Verso.

Bartelson, J. (1995) *A Genealogy of Sovereignty*, Cambridge: Cambridge University Press.

Barth, F. (ed.) (1993) *Balinese Worlds*, Chicago: University of Chicago Press.

Benedick, R. (1991) *Ozone Diplomacy: New Directions in Safeguarding the Planet*, Cambridge, MA: Harvard University Press.

Benton, L.M. and Short, J.R. (1999) *Environmental Discourse and Practice*, Oxford and Malden, MA: Blackwell.

Berger, P. and Luckmann, T. (1966) *The Social Construction of Reality: A Treatise in the Sociology of Knowledge*, 1984 edn, Harmondsworth: Penguin.

Bergman, R. (1997) 'Hazards from nuclear materials in the arctic region: do we know sufficient for assessing risks and deciding on priorities?', in D. Vidas (ed.) *Arctic Development and Environmental Challenges. Information Needs for Decision-Making and International Co-operation*, Copenhagen: Scandinavian Seminar College.

Bergman, R. and Baklanov, A. (1998) *Radioactive Sources of Main Radiological Concern*, Stockholm: Swedish Council for Planning and Coordination of Research/Swedish Defence Research Establishment.

Bernstein, R.J. (1976) *The Restructuring of Social and Political Theory*, 1979 edn, London: Methuen.

Berteig, A., Hønneland, G., Jørgensen, A.K. and Pachina, T. (1998) *Public Child Care on the Kola Peninsula: A Study of the System for Public Care of Orphans and Neglected Children on the Kola Peninsula*, Oslo: SOS Children's Villages.

Billig, M. (1987) *Arguing and Thinking: A Rhetorical Approach to Social Psychology*, Cambridge: Cambridge University Press.

Bond, A. (1996) 'The Russian copper industry and the Noril'sk joint-stock company in the mid-1990s', *Post-Soviet Geography and Economics* 37: 286–329.

Bond, A. and Levine, R.M. (2001) 'Noril'sk nickel and Russian platinum-group metals production', *Post-Soviet Geography and Economics* 42: 77–104.

Cantrill, J.G. and Oravec, C.L. (eds) (1996) *The Symbolic Earth. Discourse and Our Creation of the Environment*, Lexington, KY: University Press of Kentucky.

Churchill, R. and Ulfstein, G. (1992) *Marine Management in Disputed Areas: The Case of the Barents Sea*, London: Routledge.

Cocklin, C. (1995) 'Agriculture, society and environment: discourses on sustainability', *International Journal of Sustainable Development and World Ecology* 2: 240–256.

Cohen, S.F. (2000) *Failed Crusade: America and the Tragedy of Post-Communist Russia*, New York and London: W.W. Norton.

Dahlström, M., Eskelinen, H. and Wiberg, U. (eds) (1995) *The East–West Interface in the European North*, Uppsala: Nordisk Samhällsgeografisk Tidskrift.

Darst, R.G. (2001) *Smokestack Diplomacy: Cooperation and Conflict in East–West Environmental Politics*, Cambridge, MA and London: MIT Press.

Davies, B. and Harré, R. (1990) 'Positioning: the discursive production of selves', *Journal for the Theory of Social Behaviour* 20: 43–63.

Dellenbrant, J.Å. and Olsson, M.O. (eds) (1994) *The Barents Region. Security and Economic Development in the European North*, Umeå: CERUM.

Dellenbrant, J.Å. and Wiberg, U. (eds) (1997) *Euro-Arctic Curtains*, Umeå: CERUM.

Douglas, M. (ed.) (1982) *Essays in the Sociology of Perception*, London: Routledge and Kegan Paul.

Douglas, M. (1987) *How Institutions Think*, London: Routledge and Kegan Paul.

Dryzek, J.S. (1997) *The Politics of the Earth: Environmental Discourses*, Oxford: Oxford University Press.

Eder, K. and Kousis, M. (2000) *Environmental Politics in Southern Europe: Actors, Institutions and Discourses in a Europeanizing Society*, Dordrecht: Kluwer Academic.

FAO (1995) *Code of Conduct for Responsible Fisheries*, signed in Rome, 28 September 1995, Rome: UN Food and Agriculture Organisation.

Feyerabend, P. (1975) *Against Method: Outline of an Anarchistic Theory of Knowledge*, London: Verso.

Flikke, G. (ed.) (1998) *The Barents Region Revisited*, Oslo: Norwegian Institute of International Affairs.

Foucault, M. (1972) *The Archeology of Knowledge*, London: Tavistock.

Foucault, M. (1973) *The Order of Things*, New York: Pantheon.

Foucault, M. (1980) *Power/Knowledge: Selected Interviews and Other Writings*, New York: Pantheon.

Freestone, D. and Hey, E. (1996) *The Precautionary Principle and International Law: The Challenge of Implementation*, The Hague: Kluwer Law International.

Garcia, S.M. (1994) 'The precautionary principle: its implications in capture fisheries management', *Ocean and Coastal Management* 22: 99–125.

Gardner, R., Ostrom, E. and Walker, J. (1994) 'The nature of common-pool resource problems', *Rationality and Society* 2: 335–358.

Gibbons, M.T. (ed.) (1987) *Interpreting Politics*, Oxford: Blackwell.

Giddens, A. (1979) *Central Problems in Social Theory: Action, Structure and Contradiction in Social Analysis*, Berkeley, CA: University of California Press.

Giddens, A. (1984) *The Constitution of Society*, Cambridge: Polity Press.

Goldstein, J. (1989) 'The impact of ideas on trade policy', *International Organization* 43: 31–71.

Government of the Russian Federation (1992) *Polozheniye o federalnom ekologich-eskom fonde Rossiyskoy Federatsii*, confirmed by Presidential Decree No. 442 of 29 June 1992, Moscow: Government of the Russian Federation.

Government of the Russian Federation (1999) *O litsenzirovanii deyatelnosti po ispolzovaniyu radioaktivnykh materialov pri provedenii rabot po ispolzovaniyu atomnoy energii v oboronnykh tselyakh*, Order of the Government of the Russian Federation No. 1007 of 4 September 1999, Moscow: Government of the Russian Federation.

Haas, P. (1989) 'Do regimes matter? Epistemic communities and Mediterranean pollution control', *International Organization* 43: 377–404.

Haas, P. (1992a) 'Introduction: epistemic communities and international policy coordination', *International Organization* 46: 1–36.

Haas, P. (1992b) 'Banning chlorofluorocarbons: efforts to protect stratospheric ozone', *International Organization* 46: 187–224.

Hajer, M.A. (1995) *The Politics of Environmental Discourse: Ecological Modern-ization and the Policy Process*, Oxford: Clarendon Press.

Hall, P.A. (1989) *The Political Power of Economic Ideas: Keynesianism across Nations*, Princeton, NJ: Princeton University Press.

Hanf, K. (2000) 'The problem of long-range transport of air pollution and the acidification regime', in A. Underdal and K. Hanf (eds) *International Environ-mental Agreements and Domestic Politics: The Case of Acid Rain*, Aldershot: Ashgate.

Hansen, E. and Tønnessen, A. (1998) *Environment and Living Conditions on the Kola Peninsula*, Oslo: FAFO Institute for Applied Social Science.

Harré, R. (1993) *Social Being*, Oxford: Blackwell.

Hewison, G.J. (1996) 'The precautionary principle to fisheries management: an environmental perspective', *International Journal of Marine and Coastal Law* 11: 301–332.

Hoel, A.H. (1994) 'The Barents Sea: fisheries resources for Europe and Russia', in O.S. Stokke and O. Tunander (eds) *The Barents Region: Cooperation in Arctic Europe*, London: Sage.

Hoel, A.H., Jentoft, S. and Mikalsen, K. (1996) 'User-group participation in Norwegian fisheries management', in R.M. Meyer, C. Zhang, M.L. Windsor, B.J. McCay, L.J. Hushak and R.M. Muth (eds) *Fisheries Resource Utilization and Policy*, New Delhi and Calcutta: Oxford and IBH.

Holm, P. and Mazany, L. (1995) 'Changes in the organization of the Norwegian fishing industry', *Marine Resource Economics* 10: 299–312.

Hønneland, G. (1994) *Kultur som identitetsskaper og tilrettelegger for en funksjonell Barentsregion*, Tromsø: NORUT Social Science Research Ltd.

Hønneland, G. (1996) 'Identitet og funksjonalitet i Barentsregionen', *Internasjonal Politikk* 54: 3–32.

Hønneland, G. (1998a) 'Autonomy and regionalisation in the fisheries management of Northwestern Russia', *Marine Policy* 22: 57–65.

Hønneland, G. (1998b) 'Compliance in the Fishery Protection Zone around Svalbard', *Ocean Development and International Law* 29: 339–360.

Hønneland, G. (1998c) 'Identity formation in the Euro-Arctic Barents Region', *Cooperation and Conflict* 33: 277–297.

Hønneland, G. (1999) 'Co-operative action between fishermen and inspectors in the Svalbard Zone', *Polar Record* 35: 207–214.

Hønneland, G. (2000a) *Coercive and Discursive Compliance Mechanisms in the Management of Natural Resources: A Case Study from the Barents Sea Fisheries*, Dordrecht: Kluwer Academic.

Hønneland, G. (2000b) 'Enforcement co-operation between Norway and Russia in the Barents Sea fisheries', *Ocean Development and International Law* 31: 249–267.

Hønneland, G. (2001) 'Fisheries in the Svalbard Zone: legality, legitimacy and compliance', in A.G. Oude Elferink and D.R. Rothwell (eds) *The Law of the Sea and Polar Maritime Delimitation and Jurisdiction*, The Hague: Kluwer Law International.

Hønneland, G. and Blakkisrud, H. (eds) (2001) *Centre–Periphery Relations in Russia: The Case of the Northwestern Regions*, Aldershot: Ashgate.

Hønneland, G. and Jørgensen, A.K. (1999a) *Integration vs. Autonomy: Civil–Military Relations on the Kola Peninsula*, Aldershot: Ashgate.

Hønneland, G. and Jørgensen, A.K. (1999b) 'A cross-border perspective on a north Russian gateway', *Post-Soviet Geography and Economics* 40: 44–61.

Hønneland, G. and Jørgensen, A.K. (2003) *Implementing International Environmental Agreements in Russia*, Manchester: Manchester University Press.

Hønneland, G. and Moe, A. (2000) *Evaluation of the Norwegian Plan of Action for Nuclear Safety: Priorities, Organisation, Implementation*, Evaluation Report 7/2000, Oslo: Norwegian Ministry of Foreign Affairs.

Hønneland, G. and Moe, A. (2002) *Evaluation of the Barents Health Programme*, Lysaker: Fridtjof Nansen Institute.

Hønneland, G. and Nilssen, F. (2000) 'Comanagement in Northwest Russian fisheries', *Society and Natural Resources* 13: 635–648.

ICES (1997) *Report of the Study Group on the Precautionary Approach to Fisheries Management*, Copenhagen: International Council for the Exploration of the Sea.

ICES (1998a) *Report of the Study Group on the Precautionary Approach to Fisheries Management*, Copenhagen: International Council for the Exploration of the Sea.

ICES (1998b) *ACFM Report 1998*, Copenhagen: International Council for the Exploration of the Sea.

ICES (2000) *ACFM Report 2000*, Copenhagen: International Council for the Exploration of the Sea.

ICES (2001) *ACFM Report 2001*, Copenhagen: International Council for the Exploration of the Sea.

Jørgensen, A.K. (2001) 'The military sector: federal responsibility – regional concern', in G. Hønneland and H. Blakkisrud (eds) *Centre–Periphery Relations in Russia: The Case of the Northwestern Regions*, Aldershot: Ashgate.

Jørgensen, A.K. and Hønneland, G. (2002) *Over grensen etter kunnskap? Evaluering av 13 prosjekter innenfor satsingsområdet kompetanse og utdanning finansiert over Barentsprogrammet*, Lysaker: Fridtjof Nansen Institute.

Kaye, S.M. (2001) *International Fisheries Management*, The Hague: Kluwer Law International.

Kirkow, P. (1997) 'Transition in Russia's principal coastal gateways', *Post-Soviet Geography and Economics* 38: 296–314.

Kotov, V. and Nikitina, E. (1996) 'Russia wrestles with an old polluter', *Environment* 38: 6–11 and 32–37.

Kotov, V. and Nikitina, E. (1998a) 'Regime and enterprise: Norilsk Nickel and transboundary air pollution', in D.G. Victor, K. Raustiala and E.B. Skolnikoff (eds) *The Implementation and Effectiveness of International Environmental Commitments: Theory and Practice*, Cambridge, MA and London: MIT Press.

Kotov, V. and Nikitina, E. (1998b) 'Implementation and effectiveness of the acid rain regime in Russia', in D.G. Victor, K. Raustiala and E.B. Skolnikoff (eds) *The Implementation and Effectiveness of International Environmental Commitments: Theory and Practice*, Cambridge, MA and London: MIT Press.

Krasner, S.D. (1982) 'International regimes', *International Organization* 36: 185–205.

Kratochwil, F.V. (1990) *Rules, Norms, and Decisions: On the Conditions of Practical and Legal Reasoning in International and Domestic Affairs*, Cambridge: Cambridge University Press.

Kuhn, T.S. (1962) *The Structure of Scientific Revolutions*, 1970 edn, Chicago: University of Chicago Press.

Lash, S., Szerszynski, B. and Wynne, B. (1996) *Risk, Environment and Society: Towards a New Ecology*, London: Sage.

Litfin, K.T. (1994) *Ozone Discourses. Science and Politics in Global Environmental Cooperation*, New York: Columbia University Press.

Litfin, K.T. (1995) 'Framing science: precautionary discourse and the ozone treaties', *Millennium: Journal of International Studies* 24: 251–277.

Lønne, O.J., Sætre, R., Tikhonov, S., Gabrielsen, G.W., Loeng, H., Dahle, S. and Shevlyagin, K. (eds) (1997) *Status Report on the Marine Environment of the Barents Region: Report from the Working Group on the Marine Environment of the Barents Regions to the Joint Norwegian–Russian Commission on Environmental Co-operation*, Oslo: Ministry of the Environment.

Ministry of Defence (1996) *Declaration among the Department of Defence of the United States of America, the Royal Ministry of Defence of the Kingdom of Norway, and the Ministry of Defence of the Russian Federation, on Arctic Military Environmental Co-operation*, signed in Bergen, 26 September 1996, Oslo: Ministry of Defence.

Ministry of Fisheries (1997) *Protokoll for den 26. sesjon i Den blandede norsk-russiske fiskerikommisjon*, Oslo: Ministry of Fisheries.

Ministry of Fisheries (1998) *Protokoll for den 27. sesjon i Den blandede norsk-russiske fiskerikommisjon*, Oslo: Ministry of Fisheries.

Ministry of Fisheries (1999) *Protokoll for den 28. sesjon i Den blandede norsk-russiske fiskerikommisjon*, Oslo: Ministry of Fisheries.

Ministry of Fisheries (2000) *Protokoll for den 29. sesjon i Den blandede norsk-russiske fiskerikommisjon*, Oslo: Ministry of Fisheries.

Ministry of Foreign Affairs (1993) *Declaration on Co-operation in the Barents Euro-Arctic Region*, signed in Kirkenes, 11 January 1993, Oslo: Ministry of Foreign Affairs.

Ministry of Foreign Affairs (1995) *Plan of Action for the Implementation of Report No. 34 (1993–94) to the Storting on Nuclear Activities and Chemical Weapons in Areas Adjacent to our Northern Borders*, Oslo: Ministry of Foreign Affairs.

Ministry of Foreign Affairs (1998a) *Agreement between the Government of the Kingdom of Norway and the Government of the Russian Federation on Environmental Co-operation in Connection with the Dismantling of Russian Nuclear Powered Submarines withdrawn from the Navy's Service in the Northern Region, and the Enhancement of Nuclear and Radiation Safety*, signed in Moscow on 26

May 1998. Oslo: Ministry of Foreign Affairs.

Ministry of Foreign Affairs (1998b) *Protokoll for møte i den Felles Norsk-Russiske Kommisjon (Moskva, 29.–30. juli 1998)*, Oslo: Ministry of Foreign Affairs.

Ministry of Foreign Affairs (1999a) *Declaration on Principles Regarding a Multilateral Nuclear Environmental Programme in the Russian Federation*, signed in Bodø, 5 March 1999, Oslo: Ministry of Foreign Affairs.

Ministry of Foreign Affairs (1999b) *Referat fra 2. Møte i Felleskommisjonen under Rammeavtalen om atomsikkerhet mellom Norge og Russland, Oslo, 7.–8. juni 1999*, Oslo: Ministry of Foreign Affairs.

Ministry of Foreign Affairs (2000) *Annex to Plan of Action for Nuclear Safety Issues: List of Measures and Projects*, Oslo: Ministry of Foreign Affairs.

Moss Maritime (1998) *Collaboration Agreement (Contract for Society on Partnership) for co-operation in design, manufacturing and commissioning of four Special Railway Cars for the Transportation of Containers with Spent Nuclear Fuel between Kværner Maritime a.s., a company duly organized and existing under the laws of Norway, with its head office in Lysaker, Norway (hereinafter referred to as KMAR) and Interindustry scientific and technical Coordination Center of nuclide products 'Nuklid', a state unitary enterprise of the RF Ministry of Atomic Energy, duly organized and existing under the laws of the Russian Federation, with its head office in St. Petersburg, Russia (hereinafter referred to as ICC Nuklid)*, Article 6.3, signed 18 November 1998, Lysaker: Moss Maritime.

Murmansk Oblast (1997a) *Ob oblastnom ekologicheskom fonde*, Regional Law of Murmansk Oblast No. 60–02–ZMO of 3 June 1997, Murmansk: Murmansk Oblast.

Murmansk Oblast (1997b) 'Dogovor o razgranichenii predmetov vedeniya i polnomochiy mezhdu organami gosudarstvennoy vlasti Rossiyskoy Federatsii i organami gosudarstvennoy vlasti Murmanskoy oblasti', signed in Moscow 30 November 1997, published in *Rossiyskiye vesti*, 25 December 1997, Murmansk: Murmansk Oblast.

Murmansk Oblast (1998) *Soglasheniye mezhdu administratsiyey Murmanskoy oblasti i Ministerstvom Rossiyskoy Federatsii po atomnoy energii o vzaimnom sotrudnichestve v oblasti obrashcheniya s RAO i OYAT i razvitii atomnoy energetiki na territorii Murmanskoy oblasti*, signed in Murmansk on 5 May 1998, Murmansk: Murmansk Oblast.

Murmansk Oblast (2000) *Soglasheniye mezhdu administratsiyey Murmanskoy oblasti, Federalnymi organami ispolnitelnoy vlasti na territorii Murmanskoy oblasti, KNTS RAN, Severnym Flotom VMF MO RF i gosudarstvennymi predpriyatiyami po koordinatsii vzaimodeystviya v sfere yadernoy i radiatsionnoy bezopasnosti na territorii Murmanskoy oblasti* No. 37–2/296, signed in Murmansk on 7 March 2000, Murmansk: Murmansk Oblast.

Myerson, G. and Rydin, Y. (1996) *The Language of Environment: A New Rhetoric*, London: UCL Press.

Nakken, O. (1998) 'Past, present and future exploitation and management of marine resources in the Barents Sea and adjacent areas', *Fisheries Research* 37: 23–35.

Neumann, I. (1994) 'A region-building approach to northern Europe', *Review of International Studies* 20: 53–74.

Neumann, I.B. (2001) *Mening, materialitet, makt: En innføring i diskursanalyse*, Bergen: Fagbokforlaget.

Nilsen, T., Kudrik, I. and Nikitin, A. (1996) *The Russian Northern Fleet: Sources of Radioactive Contamination*, Bellona Report No. 2, Oslo: Bellona.

Norwegian Radiation Protection Agency (1997) *Avtale mellom Statens strålevern, Norge, og Russlands Føderale Tilsyn for kjerne- og strålingssikkerhet om teknisk samarbeid og utveksling av informasjon vedrørende sikker bruk av atomenergi*, signed in Moscow on 20 October 1997, Oslo: Norwegian Radiation Protection Agency.

OECD (1999) *Environmental Performance Reviews: Russian Federation*, Paris: Organisation for Economic Co-operation and Development, Centre for Co-operation with Non-Members.

Ostrom, E., Gardner, R. and Walker, J. (1994) *Rules, Games, and Common-Pool Resources*, Ann Arbor, MI: University of Michigan Press.

Russian Federation (1998) *Ob Isklyuchitelnoy Ekonomicheskoy Zone Rossiyskoy Federatsii*, Federal Law, adopted by the State Duma 18 November 1998 and by the Federation Council 2 December 1998, Moscow: Russian Federation.

Russian Soviet Federative Socialist Republic (1991) *Ob okhrane okruzhayushey prirodnoy sredy*, Republican Law of the Russian Soviet Federative Socialist Republic No. 2060–1 of 19 December 1991, Moscow: Russian Federative Socialist Republic.

Sawhill, S. (2000) 'Cleaning-up the Arctic's cold war legacy: nuclear waste and arctic military environmental cooperation', *Cooperation and Conflict* 35: 5–36.

Sawhill, S. and Jørgensen, A.K. (2001) *Military Nuclear Waste and International Cooperation in Northwest Russia*, Lysaker: Fridtjof Nansen Institute.

Schattschneider, E.E. (1960) *The Semi-Sovereign People*, New York: Rinehart and Winston.

Singh, E.C. and Saguirian, A. (1993) 'The Svalbard archipelago: the role of surrogate negotiators', in O.R. Young and G. Osherenko (eds) *Polar Politics. Creating International Environmental Regimes*, Ithaca, NY and London: Cornell University Press.

Skogan, J.K. (1986) *Sovjetunionens Nordflåte 1968–85*, Oslo: Norwegian Institute of International Affairs.

State Committee for Environmental Protection (1998) *Information Leaflet*, Murmansk: State Committee for Environmental Protection in Murmansk Oblast.

State Committee for Environmental Protection (2001) *Sostoyaniye prirodnoy okruzhayushey sredy Murmanskoy oblasti na Kolskom poluostrove v 2000 g.*, Murmansk: State Committee for Environmental Protection in Murmansk Oblast.

Stokke, O.S. (1995) *Fisheries Management under Pressure: A Changing Russia and the Effectiveness of the Barents Sea Regime*, Lysaker: Fridtjof Nansen Institute.

Stokke, O.S. (1998) 'Nuclear dumping in arctic seas: Russian implementation of the London Convention', in D. Victor, K. Raustiale and E.B. Skolnikoff (eds) *The Implementation and Effectiveness of International Environmental Commitments: Theory and Practice*, Cambridge, MA and London: MIT Press.

Stokke, O.S. (2000a) 'Radioactive waste in the Barents and Kara Seas: Russian implementation of the global dumping regime', in D. Vidas (ed.) *Protecting the Polar Marine Environment: Law and Policy for Pollution Prevention*, Cambridge: Cambridge University Press.

Stokke, O.S. (2000b) 'Subregional cooperation and protection of the arctic marine environment: the Barents Sea', in D. Vidas (ed.) *Protecting the Polar Marine Environment: Law and Policy for Pollution Prevention*, Cambridge: Cambridge University Press.

Stokke, O.S. and Hoel, A.H. (1991) 'Splitting the gains: political economy of the Barents Sea fisheries', *Cooperation and Conflict* 26: 49–65.

Stokke, O.S. and Tunander, O. (eds) (1994) *The Barents Region: Cooperation in Arctic Europe*, London: Sage.

Stokke, O.S., Anderson, L.G. and Mirovitskaya, N. (1999) 'The Barents Sea fisheries', in O.R. Young (ed.) *The Effectiveness of International Regimes: Causal Connections and Behavioral Mechanisms*, Cambridge, MA and London: MIT Press.

Stone, D.A. (1988) *Policy Paradox and Political Reason*, New York: HarperCollins.

Stortinget (1994) *Report No. 34 (1993–94) to the Storting on Nuclear Activities and Chemical Weapons in Areas Adjacent to our Northern Borders*, Oslo: Stortinget.

Stortinget (1997) *St meld nr 51 (1997–98) Perspektiver på norsk fiskerinæring*, Oslo: Stortinget.

Stortinget (2002) *Innst. S. nr. 107. Innstilling fra kontroll- og konstitusjonskomiteen om Riksrevisjonens undersøkelse av regjeringens gjennomføring av Handlingsplanen for atomsaker*, Oslo: Stortinget.

Tennberg, M. (2000) *Arctic Environmental Co-operation: A Study in Governmentality*, Aldershot: Ashgate.

Torgerson, D. (1990) 'Limits of the administrative mind: the problem of defining environmental problems', in R. Paehlke and D. Torgerson (eds) *Managing Leviathan: Environmental Politics and the Administrative State*, Peterborough: Broadview.

Ulfstein, G. (1995) *The Svalbard Treaty: From Terra Nullius to Norwegian Sovereignty*, Oslo: Scandinavian University Press.

United Nations (1982) 'United Nations Convention on the Law of the Sea', Montego Bay, 10 December 1982, UN Doc. A/CONF. 62/122, *International Legal Materials* 21: 1261.

United Nations (1992) 'Rio Declaration on Environment and Development', Rio de Janeiro, 16 June 1992, *International Legal Materials* 31: 874.

United Nations (1995) 'Agreement for the Implementation of the Provisions of the United Nations Convention on the Law of the Sea of 10 December 1982 Relating to the Conservation and Management of Straddling Fish Stocks and Highly Migratory Fish Stocks', New York, 4 August 1995, *International Legal Materials* 34: 1547–1580.

VanderZwaag, D. (2000) 'Land-based marine pollution and the Arctic: polarities between principles and practice', in D. Vidas (ed.) *Protecting the Polar Marine Environment: Law and Policy for Pollution Prevention*, Cambridge: Cambridge University Press.

Weinberg, A. (1972) 'Science and trans-science', *Minerva* 10: 209–222.

Wiberg, U. (1994) 'From visions to functional relationships in the Barents region', paper presented at the conference 'Trans-border regional co-operation in the Barents region', Kirkenes, 24–26 February.

Wissenburg, M., Orhan, G. and Collier, U. (1999) *European Discourses on Environmental Policy*, Aldershot: Ashgate.

Yablokov, A.V., Karasev, V.K., Rumyanstsev, V.M., Kokeev, M.E., Petrov, O.Y., Lystsov, V.N., Emel'yanenkov, A.M. and Rubtsov, P.M. (1993) *Facts and Problems Related to Radioactive Waste Disposal in Seas Adjacent to the Territory of the Russian Federation*, Moscow: Small World Publishers.

Index

Note: 'n.' after a page reference indicates the number of a note on that page.

Printed and bound by CPI Group (UK) Ltd, Croydon, CR0 4YY

01/11/2024

01782615-0019